J. A. SEIGEL

SPACE SCIENCE IN THE TWENTY-FIRST CENTURY: IMPERATIVES FOR THE DECADES 1995 TO 2015

ASTRONOMY AND ASTROPHYSICS

Task Group on Astronomy and Astrophysics
Space Science Board
Commission on Physical Sciences, Mathematics, and Resources
National Research Council

NATIONAL ACADEMY PRESS
Washington, D.C. 1988

National Academy Press • 2101 Constitution Avenue, N.W. • Washington, D. C. 20418

NOTICE: The project that is the subject of this report was approved by the Governing Board of the National Research Council, whose members are drawn from the councils of the National Academy of Sciences, the National Academy of Engineering, and the Institute of Medicine. The members of the committee responsible for the report were chosen for their special competences and with regard for appropriate balance.

This report has been reviewed by a group other than the authors according to procedures approved by a Report Review Committee consisting of members of the National Academy of Sciences, the National Academy of Engineering, and the Institute of Medicine.

The National Academy of Sciences is a private, nonprofit, self-perpetuating society of distinguished scholars engaged in scientific and engineering research, dedicated to the furtherance of science and technology and to their use for the general welfare. Upon the authority of the charter granted to it by the Congress in 1863, the Academy has a mandate that requires it to advise the federal government on scientific and technical matters. Dr. Frank Press is president of the National Academy of Sciences.

The National Academy of Engineering was established in 1964, under the charter of the National Academy of Sciences, as a parallel organization of outstanding engineers. It is autonomous in its administration and in the selection of its members, sharing with the National Academy of Sciences the responsibility for advising the federal government. The National Academy of Engineering also sponsors engineering programs aimed at meeting national needs, encourages education and research, and recognizes the superior achievements of engineers. Dr. Robert M. White is president of the National Academy of Engineering.

The Institute of Medicine was established in 1970 by the National Academy of Sciences to secure the services of eminent members of appropriate professions in the examination of policy matters pertaining to the health of the public. The Institute acts under the responsibility given to the National Academy of Sciences by its congressional charter to be an adviser to the federal government and, upon its own initiative, to identify issues of medical care, research, and education. Dr. Samuel O. Thier is president of the Institute of Medicine.

The National Research Council was organized by the National Academy of Sciences in 1916 to associate the broad community of science and technology with the Academy's purposes of furthering knowledge and advising the federal government. Functioning in accordance with general policies determined by the Academy, the Council has become the principal operating agency of both the National Academy of Sciences and the National Academy of Engineering in providing services to the government, the public, and the scientific and engineering communities. The Council is administered jointly by both Academies and the Institute of Medicine. Dr. Frank Press and Dr. Robert M. White are chairman and vice chairman, respectively, of the National Research Council.

Support for this project was provided by Contract NASW 3482 between the National Academy of Sciences and the National Aeronautics and Space Administration.

Library of Congress Catalog Card Number 87-43333

ISBN 0-309-03875-8

Printed in the United States of America

TASK GROUP ON ASTRONOMY AND ASTROPHYSICS

Bernard Burke, Massachusetts Institute of Technology, Chairman
James Roger Angel, University of Arizona
Jacques Beckers, NOAO Advanced Development Program
Andrea Dupree, Center for Astrophysics
Carl E. Fichtel, NASA Goddard Space Flight Center
George Field, Center for Astrophysics
Riccardo Giacconi, Space Telescope Science Institute
Jonathan Grindlay, Center for Astrophysics
Martin Harwit, Cornell University
Frank Low, University of Arizona
Frank McDonald, NASA Headquarters
Dietrich Muller, University of Chicago
Minoru Oda, ISAS
Klaus Pinkau, Max-Planck Institute for Plasma Physics
Kenneth A. Pounds, University of Leicester
Irwin Shapiro, Center for Astrophysics
Susan Wyckoff, Arizona State University

Richard C. Hart, *Staff Officer*
Carmela J. Chamberlain, *Secretary*

STEERING GROUP

Thomas M. Donahue, University of Michigan, Chairman
Don L. Anderson, California Institute of Technology
D. James Baker, Joint Oceanographic Institutions, Inc.
Robert W. Berliner, Pew Scholars Program, Yale University
Bernard F. Burke, Massachusetts Institute of Technology
A. G. W. Cameron, Harvard College Observatory
George B. Field, Center for Astrophysics, Harvard University
Herbert Friedman, Naval Research Laboratory
Donald M. Hunten, University of Arizona
Francis S. Johnson, University of Texas at Dallas
Robert Kretsinger, University of Virginia
Stamatios M. Krimigis, Applied Physics Laboratory
Eugene H. Levy, University of Arizona
Frank B. McDonald, NASA Headquarters
John E. Naugle, Chevy Chase, Maryland
Joseph M. Reynolds, The Louisiana State University
Frederick L. Scarf, TRW Systems Park
Scott N. Swisher, Michigan State University
David A. Usher, Cornell University
James A. Van Allen, University of Iowa
Rainer Weiss, Massachusetts Institute of Technology

Dean P. Kastel, *Study Director*
Ceres M. Rangos, *Secretary*

SPACE SCIENCE BOARD

Thomas M. Donahue, University of Michigan, Chairman
Philip H. Abelson, American Association for the Advancement of Science
Roger D. Blandford, California Institute of Technology
Larry W. Esposito, University of Colorado
Jonathan E. Grindlay, Center for Astrophysics
Donald N. B. Hall, University of Hawaii
Andrew P. Ingersoll, California Institute of Technology
William M. Kaula, NOAA
Harold P. Klein, The University of Santa Clara
John W. Leibacher, National Solar Observatory
Michael Mendillo, Boston University
Robert O. Pepin, University of Minnesota
Roger J. Phillips, Southern Methodist University
David M. Raup, University of Chicago
Christopher T. Russell, University of California, Los Angeles
Blair D. Savage, University of Wisconsin
John A. Simpson, Enrico Fermi Institute, University of Chicago
George L. Siscoe, University of California, Los Angeles
L. Dennis Smith, Purdue University
Darrell F. Strobel, Johns Hopkins University
Byron D. Tapley, University of Texas at Austin

Dean P. Kastel, *Staff Director*
Ceres M. Rangos, *Secretary*

COMMISSION ON PHYSICAL SCIENCES, MATHEMATICS, AND RESOURCES

Norman Hackerman, Robert A. Welch Foundation, Chairman
George F. Carrier, Harvard University
Dean E. Eastman, IBM Corporation
Marye Anne Fox, University of Texas
Gerhart Friedlander, Brookhaven National Laboratory
Lawrence W. Funkhouser, Chevron Corporation (retired)
Phillip A. Griffiths, Duke University
J. Ross Macdonald, University of North Carolina, Chapel Hill
Charles J. Mankin, Oklahoma Geological Survey
Perry L. McCarty, Stanford University
Jack E. Oliver, Cornell University
Jeremiah P. Ostriker, Princeton University Observatory
William D. Phillips, Mallinckrodt, Inc.
Denis J. Prager, MacArthur Foundation
David M. Raup, University of Chicago
Richard J. Reed, University of Washington
Robert E. Sievers, University of Colorado
Larry L. Smarr, National Center for Supercomputing Applications
Edward C. Stone, Jr., California Institute of Technology
Karl K. Turekian, Yale University
George W. Wetherill, Carnegie Institution of Washington
Irving Wladawsky-Berger, IBM Corporation

Raphael G. Kasper, *Executive Director*
Lawrence E. McCray, *Associate Executive Director*

Foreword

Early in 1984, NASA asked the Space Science Board to undertake a study to determine the principal scientific issues that the disciplines of space science would face during the period from about 1995 to 2015. This request was made partly because NASA expected the Space Station to become available at the beginning of this period, and partly because the missions needed to implement research strategies previously developed by the various committees of the board should have been launched or their development under way by that time. A two-year study was called for. To carry out the study the board put together task groups in earth sciences, planetary and lunar exploration, solar system space physics, astronomy and astrophysics, fundamental physics and chemistry (relativistic gravitation and microgravity sciences), and life sciences. Responsibility for the study was vested in a steering group whose members consisted of the task group chairmen plus other senior representatives of the space science disciplines. To the board's good fortune, distinguished scientists from many countries other than the United States participated in this study.

The findings of the study are published in seven volumes: six task group reports, of which this volume is one, and an overview report of the steering group. I commend this and all the other task group reports to the reader for an understanding of the challenges that confront the space sciences and the insights they promise for the next century. The official recommendations of the study are those to be found in the steering group's overview.

Thomas M. Donahue, Chairman
Space Science Board

Contents

1. INTRODUCTION AND SUMMARY 1
   The Importance of Advanced Instrumentation
   in Astrophysical Progress, 3
   Astrophysical Goals, 4
   The Next Generation of Powerful Observatories
   in Space, 6

2. SCIENCE OBJECTIVES 8
   Introduction, 8
   Basic Astrophysical Questions, 9
   Relationship to Other Disciplines of Space Science, 17

3. ASTRONOMY AND ASTROPHYSICS IN 1995:
   EXPECTED STATUS 19
   Overview, 19
   Radio Astronomy, 20
   Infrared and Submillimeter Astronomy, 22
   Ultraviolet and Optical Wavelengths, 24
   X-ray Astronomy, 25
   Gamma-Ray Astronomy, 27
   Cosmic-Ray Astrophysics, 28
   Gravitational Physics, 31
   NASA Operational Status, 33
   International Programs, 33

4. NEW INITIATIVES 36
    The Way Forward, 36
    High-Resolution Interferometry, 38
    High-Throughput Instruments, 46
    Very High Throughput Facility (VHTF), 56
    Hard X-ray Imaging Facility (HXIF), 58
    Gamma-Ray Astronomy, 60
    Cosmic-Ray Research, 62

5. PRACTICAL CONSIDERATIONS 67
    Budgetary Requirements, 67
    International Collaboration, 68
    Cost-to-Weight Ratio, 69
    Management and Operations, 69
    Coordinated Facilities, 70
    Scientific Instruments Technology, 70

1
Introduction and Summary

The past 60 years have brought about revolutionary changes in our understanding of the universe. For the first time we have been able to measure its size and age, to look back toward its birth in the very distant past, and to understand its present appearance and past evolution in terms of the laws of physics and chemistry discovered on Earth.

The intellectual impact of astronomy, in union with physics, continues today. Recently, we have found that our improved theoretical understanding of elementary particles, the fundamental constituents of matter, allows us to ask deep questions about the nature of the universe that we could not even formulate a short time ago. Discovery of the underlying relationship between the large-scale properties of the universe and the microscopic laws of physics would be a new triumph of human thought, comparable in impact with the Newtonian synthesis or Einstein's theory of general relativity. The extraordinary richness of astronomical phenomena ensures that the age of discovery is not behind us and that "there are more things in Heaven and Earth, Horatio, than are dreamt of in your philosophy."

Even though the details of astronomical knowledge change as new discoveries are made, there are three major themes that have guided the astronomer's quest for knowledge for several decades.

The same themes, in the judgment of the task group, will continue to express the aims of astrophysics for the foreseeable future. These are to understand the following:

- the origin of the universe;
- the laws of physics governing the universe; and
- the birth of stars and planets and the advent of life.

These goals are vast in scope, challenging in their complexity, and profound in their implications.

The experience of the past 30 years clearly shows that observations across the electromagnetic spectrum from gamma rays to radio waves, as well as detection of cosmic-ray particles, are required to study the enormous variety of physical conditions found in the universe. The use of space observations has already revealed the existence of matter in previously unknown forms. Matter at densities one million billion times the density of water has been found to exist in stars at the end of their lives when they contract to become white dwarfs and neutron stars. Similarly, gases at temperatures of tens of millions of degrees fill the space between galaxies and, despite far lower densities, contain as much mass as that in all stars and galaxies together.

The next 30 years can be expected to present exciting new opportunities for fundamental discoveries through space-based observatories operating at all wavelengths. The size of these facilities will be comparable with the present scale of ground-based observatories. Major new instruments will permit us to observe at sensitivities that are orders of magnitude greater than any currently available.

The program of new initiatives for the era 1995 to 2015 focuses on improvements in capabilities in two areas: higher angular resolution and greater collecting area.

Over the past century, astronomy has forged increasingly strong links to all other branches of science. There are clear connections between the scientific objectives of the terrestrial, planetary, solar, and astrophysics programs. The search for life in other solar systems is a common goal; the prospects of detecting gravitational radiation have importance for physics as well as astrophysics; the study of the birth and chemical evolution of stars and planetary systems has a direct connection not only to the origins of life but also to the fundamental state of matter as it existed in the first few moments after the universe began its explosive expansion.

Astrophysics, like most of the basic sciences, is international by nature. The United States has held a leading position in this field, innovating in using new technology to construct highly sensitive instruments both on the ground and in space. In turn, the success of our scientists has inspired the best scientific and engineering talents to join the space effort. Cooperation with scientists of other nations has also been a considerable source of strength. The task group's plan for the era 1995 to 2015 recognizes that international cooperation will be crucial for optimizing the scientific return.

THE IMPORTANCE OF ADVANCED INSTRUMENTATION IN ASTROPHYSICAL PROGRESS

Modern astronomy traces its beginnings to 1609 when Galileo first looked at the heavens with a spyglass—an instrument that had been invented only the year before. That one invention drastically changed both the direction and the pace at which astronomy was to be conducted. Increasingly, astronomers came to utilize advanced observing techniques in all accessible parts of the spectrum. The most striking recent advances have depended on instruments designed to detect radio waves, infrared emissions, and x rays as well as the even more energetic gamma rays. These instruments have permitted the discovery of quasars, pulsars, neutron stars, whole galaxies that emit the bulk of their radiation as x rays, and other galaxies that emit most powerfully at infrared wavelengths. They have even let us "hear" the ubiquitous radio background radiation reaching us from the most distant portions of the universe ever observed—attesting to an intensely hot, explosive origin of the cosmos.

When we examine the history of technological advances for monitoring all wavelengths of radiation, we see a rapidly expanding observational capability in the post-World War II era. In the wake of this expanding front of technical competence we see new discoveries crystallizing with amazing rapidity. Astronomers first discovered quasars at radio wavelengths around 1963 through the use of instrumentation unavailable just four years earlier. X-ray stars and galaxies, discovered in 1962 and 1966, respectively, were first detected with instrumentation not yet invented in 1959. Novel, more powerful instruments were crucial to these discoveries.

Recognizing the importance of new and more powerful observing capabilities, the National Aeronautics and Space Administration has planned a family of observatories, to be launched in the new few years. Under this Great Observatories in Space program, NASA expects to launch the Hubble Space Telescope (HST), the Gamma-Ray Observatory (GRO), the Advanced X-ray Astrophysics Facility (AXAF), and the Space Infrared Telescope Facility (SIRTF). These long-lived, orbiting observatories should be active throughout the 1990s. The extensive wavelength coverage that they will provide spans most of the spectrum accessible solely from space. Their sensitivity will be far higher than any available to date. Their spectral resolving power will advance instrumental capabilities into realms unreachable before. With such powerful technological innovations, this family of great observatories should continue to uncover new phenomena whose significance will match the greatest discoveries of past decades.

Experience shows that we have been most successful in explaining complex cosmic phenomena when we can observe across the entire wavelength range. Only in that way can a variety of different processes underlying a given phenomenon be distinguished from each other. Each process typically has distinctive signatures at different wavelengths: When we observe the Milky Way at optical wavelengths, we see a steady stream of light from stars. Infrared telescopes sense the heat emitted by interstellar dust and permit us to trace the distribution of dust clouds throughout our galaxy. X-ray telescopes, in contrast, register x-ray pulses emitted by compact binary stars, while radio telescopes present us with a map showing the distribution of hydrogen clouds pervading the interstellar spaces. Finally, a gamma-ray telescope scanning the Milky Way would observe occasional powerful bursts of gamma rays from unpredictable directions at intervals of a few weeks. We do not yet understand the cause of these bursts, which last only a few seconds.

ASTROPHYSICAL GOALS

The time scale of the universe is best thought of as being governed by a peculiar kind of clock—the logarithm clock. Like a stopwatch, its face is divided by tick marks into 60 intervals. But, instead of going around at a steady pace, the pointer runs 10 times slower at each successive tick mark that it passes. It starts off at a fantastic whirl. After passing the first 2 tick marks, it has already slowed by a factor of 100. By the time 10 tick marks have been passed, it is moving 10 billion times more slowly. But its initial speed is so great that 43 tick marks have to be passed before the hand moves one tick in 1 second. The next tick mark thereafter is reached 10 seconds later, the following tick takes a 100 seconds, the next 1000 seconds, and so on. The formation of the Earth and life as we know it dates to the fifty-ninth tick mark. The pointer today points somewhere between the fifty-ninth and sixtieth marks. It will take tens of billions of years before the sixtieth tick mark is reached. This clock can help explain the importance of the earliest instant in the life of the universe.

As our clock advances, the universe expands. With every second tick mark passed, the universe expands its dimensions by a factor of 10. Passage of 10 ticks of time on the clock increases the dimensions of the universe one hundred-thousandfold. Simultaneously, the temperature of the universe—the energy of the particles that it contains—drops a similar amount, one hundred-thousandfold for every 10 ticks of the clock. The importance of this temperature drop is that the nature of physical interactions depends strongly on the energy of the interacting radiation or matter. A change in particle energy by a factor of 10 can lead to dramatic changes in the types of particles that the universe contains, in the interactions that these particles exhibit, and in the laws of physics that such matter obeys.

To understand how the universe today evolved from its initial state, we must trace back to these earliest times. In part, we can do this by using powerful telescopes that look far out into the universe. Optical and infrared observatories in space, like HST and SIRTF, will look back in time to the era when the galaxies first formed. They will be able to analyze, with their powerful spectrometers, what the chemical constituents of the universe were at that time— whether hydrogen and helium were the only constituents present or whether some lithium and deuterium also existed. Measuring the abundances of these rare substances at those early times would provide us with a sensitive determination of the density of ordinary matter in the universe at the time when helium nuclei were formed. By inference, it would also determine the nuclear matter density in the universe today.

Determining the present density of nuclear matter is important for understanding the fate of the universe. The masses computed for galaxies from observations of their rotations and mutual attractions systematically exceed the total mass of stars, gas, and dust that we can observe in these galaxies. So striking is this discrepancy that we now talk of "dark matter"—an unknown form of matter perceived only through its gravitational attraction, otherwise invisible.

With AXAF we will be able to probe the universe to an even greater depth. The distribution of gases at temperatures of millions of degrees in the halos that surround some galaxies can be used as a sensitive probe of the distribution of the gravitational force field and thereby also of the distribution of mass. HST and SIRTF will also be able to examine this mass distribution. They will look, respectively, at the gravitational pull exerted on faint populations of stars in the outer reaches of galaxies and at the populations of Jupiter-sized objects evident only through the infrared radiation they emit.

AXAF and GRO may also be able to discern the x-ray and gamma-ray emission from the earliest quasars formed. We may be able to learn whether the first large condensations to take shape were quasars, individual stars, or galaxies.

THE NEXT GENERATION OF POWERFUL OBSERVATORIES IN SPACE

While the family of great observatories will undoubtedly provide substantial advances in our understanding of the universe, the task group foresees still further instrumental advances to help resolve a number of fundamental problems.

We need to understand the formation of the solar system and of the planets. We need to learn how prevalent planetary systems are in the galaxy. To date, we know of no other system that much resembles ours. But recent infrared observations by the Infrared Astronomical Satellite (IRAS—a joint U.S.-Dutch-British project) and from ground-based observatories have provided evidence of some stars surrounded by disks of dust and of others with companions not much more massive than the planet Jupiter.

While SIRTF and HST will provide us with information on preplanetary disks surrounding nearby stars, and will even be able to detect massive planets surrounding them, large interferometers, or an extremely large single mirror, placed in earth orbit toward the end of the century, could allow detailed investigations of such planets. These observations could open the way to a search for life on other planets.

Such interferometers—high-spatial-resolution instruments—would also permit the probing of the central engines of quasars, the most luminous objects so far discovered in the universe. Quasars eject masses of material at velocities that at first sight appear to exceed the speed of light. Their brightness can vary suddenly from hour to hour, indicating an enormously compact central energy supply that powers these enigmatic sources. Both optical and infrared interferometers of the future could achieve angular resolutions measured in microarcseconds—a resolution that would allow someone on Earth to distinguish the features of a man standing on the Moon.

We will also be able to launch massive cosmic-ray detectors, for a detailed analysis of this most energetic form of galactic matter, possibly even searching for antinucleons that would herald the possibility of a symmetric universe in which galaxies of ordinary matter and antigalaxies consisting entirely of antimatter could coexist.

2
Science Objectives

INTRODUCTION

Astronomy encompasses an enormous range of interesting and fundamental questions. The universe exhibits regions far more vacant than the best vacuum that we could hope to generate in the laboratory, and compressed matter so dense that a thimbleful would weigh a billion tons. Temperatures, pressures, magnetic field strengths, and radiation densities range similarly across extremes. A practical consequence of this great variety of conditions is a huge range of cosmic sizes, of velocities, and of time scales on which dynamic processes take place. Three decades ago, it was still possible to think of a universe in which changes occurred ponderously, if at all, over the aeons. Now we detect compact neutron stars whirling about their axes 1000 times a second, emitting streams of pulses at millisecond intervals and there are processes that take place on every other conceivable time scale between milliseconds and billions of years. These same neutron stars are no more than a few kilometers across, and there is accumulating evidence for the existence of stellar black holes that would be another 10 times smaller. At the other extreme we see radio jets that span an entire cluster of galaxies. The relativistic energies of such jets are the handiwork of electrons and protons traveling along thin filaments at close to the speed of light. These particles, which stem from an unknown source, eventually are spent in gigantic billowing clouds known through their powerful radio emission.

Closer to home, and of special interest, are planetary processes. Here the history of a system can be read in its chemical composition, and we may hope to trace the origins of life and determine the prevalence (or absence) of biological systems elsewhere in the solar system, in our galaxy, and potentially in the universe.

To study all these different conditions and construct an understandable picture, we require an ability to detect the very small as well as the very large. To probe the great variety of dynamical processes, we need to be able to measure both extremely high and very low velocities; to recognize different chemical species, we need to be able to unravel their spectral signatures. These requirements dictate that the instruments that we will need must cover the widest range of photon energies—from the extreme gamma-ray portion of the spectrum to the longest radio waves. In each of these ranges we will require large, sensitive instruments that can follow rapid time variations. Higher angular resolution, as well, is needed to focus on extremely compact sources. Such compact sources will also need to be studied both at high spectral resolution and across the widest possible dynamic range to reveal internal motions and chemical composition.

BASIC ASTROPHYSICAL QUESTIONS

In identifying scientific objectives for astronomical research in the period 1995 to 2015, the Task Group on Astronomy and Astrophysics accepts that there will be major developments in astronomy and related sciences between now and then. Priorities will necessarily have to change, and entirely new projects may evolve. However, the study *Astronomy and Astrophysics for the 1980s*, conducted by the Astronomy Survey Committee of the National Research Council, enunciated a number of basic questions that are as relevant today as when they were published in 1982. The task group believes that these will remain compelling questions for the foreseeable future as well. The fundamental astrophysical issues fall into three major groups (a fourth topic—solar and stellar activity—is discussed by the Task Group on Solar and Space

Physics in the companion volume *Space Science in the Twenty-First Century: Solar and Space Physics*):

- The early universe, unidentified matter, and the origin of galaxies.
- The physics of collapse and the physics of strong fields.
- The formation of stars, planets, and life.

In what follows, the task group has not attempted to make predictions about the discoveries possible in each area before 1995. The task group points out, however, the relevance of major NASA missions to be initiated before that time.

The Early Universe, Unidentified Matter, and the Origin of Galaxies

"Big bang" cosmology—the theory espousing the explosive origin of the universe—met a crucial experimental test with the 1965 discovery of the cosmic blackbody background radiation. That measurement yielded a ratio of photons to protons in the universe and permitted calculation of the abundances of light elements produced in the first few minutes after the big bang. Studies of these abundances have since led to the conclusion that the density of normal nuclear matter in the universe is about 10 percent of that expected of a closed universe in fair agreement with dynamical constraints on the total amount of its mass.

Grand unified theories of elementary particles postulate that the universe underwent a transition from a symmetric to an asymmetric phase 10^{-35} s after the big bang, when the temperature was 10^{28} K. A rapid inflation of the universe then resulted in a present total density of matter almost exactly that of a closed universe according to theory. The density we actually observe for matter, however, falls 10 times short of that predicted density. Such theories thus imply that 90 percent of the mass of the universe is in unidentified form; if this mass is nonbaryonic, it could take the form of massive neutrinos or even more exotic new particles like photinos or axions. This would have important implications for the origin of galaxies, which must have formed in the gravitational collapse of density perturbations in the cosmically expanding matter. As the universe evolved, most of the "normal" matter has, theoretically, been compressed by gravitation to form the stars and galaxies that we observe; these processes would be profoundly influenced by the presence of the greater bulk of matter in unidentified form.

Astronomers began to realize some years ago that there is some unidentified form of mass present in the universe. Galaxies have since been shown to be embedded in giant halos composed of material whose presence can be inferred from its gravitational effects, but which has not yet been detected directly at radio, infrared, optical, or x-ray wavelengths. The matter in such halos appears to make up 5 or 10 percent of the critical cosmic density, but the data are not accurate enough to decide whether this unidentified mass could be in the form of very faint stars or whether additional matter of novel form is required.

Density estimates based on the dynamics of clusters of galaxies do not exceed 20 percent of the closure density that is calculated for a closed universe. Thus, these estimates are apparently not consistent with an inflationary universe based on grand unified theories. On the other hand, almost all visible galaxies are observed to lie within clusters or filamentary structures that occupy a small fraction of the volume of the universe. The rest of space seems to be largely void; but these voids in the distribution of galaxies may yet contain large amounts of hidden matter not counted in the analyses of clusters and superclusters—enough, in fact, to make up the 100 percent of closure density predicted by grand unified theories. Current observations have not resolved this point.

The 1982 report of the Astronomy Survey Committee recommended a number of facilities that promise to contribute to our understanding of the unaccounted mass. HST will determine the distribution of hidden mass in galactic halos by determining the motions of globular clusters. AXAF would permit a search for the x rays emitted by the hot envelopes of faint stars in galactic halos. SIRTF will search for cold matter, such as brown dwarfs— substellar bodies emitting energy by virtue of a slow contraction that liberates gravitational potential energy. This search for cold matter by SIRTF will set limits on the baryonic composition of the dark matter component in our galaxy. AXAF can also exploit the recent discovery of hot gas in the halos of some elliptical galaxies to determine the halo mass through its gravitational effects.

The NASA instruments that will be operational or under construction by 1995 have extraordinary sensitivity at wavelengths ranging from the x-ray to the infrared region. They will thus permit us to look across large distances, far back in time to the epoch of galaxy formation, when the universe was probably only 10 or 20 percent of its current size. Theorists are currently developing computer simulations of galaxy formation in a universe dominated by various types of unidentified matter. Such models have already demonstrated that if the unidentified matter is composed of massive neutrinos, then the large-scale structures such as clusters and superclusters should be more highly developed than galaxies themselves are observed to be. Hence, current attention is focused upon other particles that remain consistent with grand unified theories but are more exotic than neutrinos. The evidence for exotic particles may come from direct measurements of their annihilation products such as proton/antiproton pairs and gamma-ray photons. By 1995, it is expected that theoretical models of galaxy formation will be well-developed, and it will be possible to compare observations of distant matter in the remote past with the predictions of particle physics. SIRTF is essential for this work, since light emitted by primordial galaxies is heavily shifted into the infrared.

A puzzling feature of the universe is the asymmetry of the matter/antimatter balance within the galaxy in view of the symmetry of the laws of nature. We have no clear-cut understanding of mechanisms that would have led an initially hot, explosive universe to favor matter over antimatter, though several symmetry-breaking mechanisms can be postulated. Whether the universe on its largest scales preserves the matter/antimatter symmetry is, therefore, a question of great interest. If matter and antimatter exist in proximity, we may expect annihilation to take place at the interface where gases from these regions mix. The annihilation radiation has its own spectral signature, readily identified, and we may expect to find this annihilation radiation, if it exists. Also, antimatter of extragalactic origin might be detected with magnetic spectrometers that could analyze the cosmic particle radiation with high sensitivity.

The Physics of Collapse and Strong Fields

The discovery of radio pulsars, neutron stars, and candidate black holes in x-ray binary stellar systems demonstrated that some stars collapse beyond the density of atomic nuclei. Theoretical models indicate that the collapse occurs within a fraction of a second when the core of a star becomes unstable; the bounce of an imploding stellar envelope off the just-formed neutron-star core is sufficient to eject the stellar envelope explosively in a supernova eruption. Such events are important in the evolution of our galaxy because they are the principal source of turbulence and heat in the interstellar medium. They account for the acceleration of cosmic rays, and create the heavy elements that later become incorporated into stars and planets like those of our solar system. These elements, which are believed to originate by such explosive nucleosynthesis, can be detected through their characteristic gamma-like decay.

Neutron stars represent a unique state of matter of exceptionally high density. In spite of their high temperatures, their interiors are superfluid and are capable of sustaining magnetic fields up to 10^{12} Gauss. Observations of their rotational properties yield information about their interior structure. Further, the radiation accompanying accretion of matter from companion stars is teaching us much about the behavior of matter in strong gravitational and magnetic fields. X-ray burst sources, for instance, are believed to be neutron stars on which a sufficient layer of matter has accumulated from a companion star to initiate a thermonuclear explosion. Recent observations of the x-ray spectrum during the decay phase of such a burst in the source 4U1636-53 indicate the presence of an absorption line. If attributed to iron (as seems probable), this spectral line is redshifted by 40 percent from the spectrum of iron observed in a laboratory. If this represents the gravitational red shift at the surface of a neutron star, then such stars are both smaller and denser than indicated by models based on general relativity and the equation of state of nuclear matter. If confirmed, this interpretation may teach us something quite unexpected about particle physics or, conceivably, about gravitation.

Some collapsing cores do not stop at the neutron star stage but plunge on toward a black hole singularity. It is of great interest to learn about the black hole state. Astronomers have identified several objects believed to be stellar black holes in orbit around a normal stellar companion. With the aid of computer simulations, theorists are now studying the accretion process near rotating black holes. They are finding that these relativistic flows exhibit novel features that could help us to understand temporal variability in the emission from black hole candidates. We anticipate that more sophisticated theoretical models will yield new insight that may tell us whether individual sources are, in fact, black holes.

Before 1995, relativistic jets will be observed in greater detail by the Very Long Baseline Array (VLBA), now under construction. Current observations have demonstrated structure on the microarcsecond scale. In several jets there are condensations that appear to be moving apart faster than the speed of light. The VLBA, with 200-μarcsec resolution, will pursue this finding with greater sensitivity and angular resolution, and the QUASAT mission will extend resolution still further, probing more deeply into the jet formation site. Much more will remain to be done, since angular resolution of a few microarcseconds will resolve details about 3 light-days across in the quasar 3C 273, and only 2 light-hours at the nucleus of the active galaxy M87. Because the Schwarzschild radius of a black hole whose mass is 10^9 times that of the Sun is about 3 light-hours, such observations will permit study of events less than 20 Schwarzschild radii from the black hole (the presumed diameter of the accretion disk), provided that the opacity is not too great. Such observations will be of extraordinary interest in testing the black hole model of active galactic nuclei and in studying the novel relativistic effects expected in the vicinity of black holes. Both the Long-Baseline Optical Space Interferometer (LBOSI) and the large space radio array (ASTROARRAY—see Chapter 4) aim at achieving 1-μarcsec resolution at optical wavelengths. We should be able to probe phenomena in active galactic nuclei at distances even closer to the black hole (light-hours) than the size of the light-emitting region (light-days) indicated by the time scales involved in light variations. Already, x-ray observations have demonstrated strong variability on a time scale of an hour or less in active galactic nuclei (AGN) of a wide range of luminosities. SIRTF will also play a vital role in the study of these superenergetic phenomena.

Gamma radiation also is a tracer for highly relativistic particles—cosmic rays—both in the galaxy and beyond. In our galaxy the gamma rays emanate from a number of discrete sources, such as the Crab and Vela pulsars, but they also emanate from dense molecular clouds. This diffuse gamma flux indicates the presence of highly energetic particles that interact with matter in these clouds. In ways we do not yet understand, this gamma flux could reflect mechanisms that play a significant role in the formation of stars. It could, for example, indicate destruction of magnetic fields that have to be eliminated in order for protostellar matter to collapse to form stars.

The mystery of the origin of cosmic rays is only partially solved at present. Acceleration of particles to relativistic energies occurs at a variety of scales: in the solar system in solar flares and planetary magnetospheres, and in the galaxy probably in stellar winds and supernova-driven shock fronts. Recent observations have indicated that binary x-ray sources containing compact objects such as Cyg X-3 appear to be powerful particle accelerators at extremely high energies. To sustain the average luminosity in cosmic rays of our galaxy (about 10^{41} ergs/s), a substantial energy source is required. If supernovae are responsible, almost 10 percent of their total energy output would be required to account for the cosmic-ray luminosity. We also know through observations of secondary photons in the radio, and sometimes in the gamma-ray region, that powerful particle acceleration must occur in almost all classes of extragalactic objects. In fact, the nonthermal radio spectra of most bright spiral galaxies seem to have almost identical slope, indicating that some common, but as yet poorly understood, mechanism governs the acceleration and propagation of the parent electrons in these galaxies.

Cosmic rays from our galaxy can be directly observed with detectors in space. At their very highest energies, they can be observed indirectly in ground-based air-shower installations. Most spaceborne gamma-ray detectors have been relatively small in size and therefore restricted to studies at low energies (below about 1 GeV) where interactions with the solar wind ("solar modulation") change the composition and energy spectra of the radiations reaching us from the galaxy. Radioactive dating, using the ^{10}Be isotope, has shown that cosmic rays are a relatively young sample of galactic matter, with an age of about 10^7 years. Yet the elemental abundance distribution of cosmic rays seems to be quite similar to that of the much older solar system. However, subtle but characteristic differences in the isotopic abundances reflect a different nucleosynthesis history of the matter from which cosmic rays are accelerated. This may constitute evidence for the continuous chemical evolution of our galaxy, but much more sensitive measurements are required before this interesting question can be resolved.

First information on the composition of the extremely rare ultraheavy cosmic rays, up to uranium, came from detectors on the HEAO-3 and Ariel-6 spacecraft. Measurements on the Space Shuttle are providing the energy spectra of individual nuclear species to much higher energies (approximately 10^{13} eV) than previously accessible and thus are expected to help reveal the character of the galactic acceleration mechanism.

Cosmic-ray measurements could potentially also lead to major changes in our understanding of fundamental physics and the formation of the universe if they detected exotic particles, such as magnetic monopoles, superheavy nuclei, or primordial antimatter.

The Formation of Stars, Planets, and Life

Our own solar system was formed 4.5 billion years ago, probably through the gravitational contraction of a fragment of an interstellar molecular cloud. With radio and infrared telescopes (IRAS is a notable recent example) we have the opportunity to observe similar processes at work today. Large numbers of active regions in molecular clouds appear to be candidates for star formation. Clouds of small particles have been discovered around several bright stars. These could conceivably represent a stage in the process of planet formation. Astronomers are already studying such events, but need higher spectral and angular resolution to sort out the complex gas dynamics. SIRTF will be able to infer the chemical composition, temperature, and density of gas flows involved in star formation regions. However, the angular resolution will be relatively limited (6 arcsec at 30-μm wavelength for SIRTF), so the images of such regions will be fuzzy. The Large Deployable Reflector (LDR), described in Chapter 5, will attain angular and spectral resolution that will permit study of details as small as 50 AU, or somewhat smaller than the diameter of the solar system, in the nearest star-forming region.

We have not yet been able to demonstrate the existence of a single planet beyond our solar system. Our instruments have so far been unable to detect any of the manifestations of such bodies: reflected light, infrared reradiation of stellar light, wobble on the plane of the sky, or periodic radial velocity variation. It should be possible in the 1990s to undertake a meaningful search for extrasolar planets by using special high-resolution spectrographs to look for the radial velocity variations in a parent star produced by planetary gravitational pulls. Space-based astrometric instruments may also be used to look for the corresponding periodic wobble of the star position.

Planetary systems may well be a common phenomenon around

late-type dwarf stars like the Sun—roughly 50 of which lie within 10 parsecs distance. Whether they are organized like our solar system into large outer planets, rich in light elements, and smaller inner planets that have high concentrations of the heavy elements is more conjectural, but plausible. Both theoretical analysis and the example of our own solar system support this view. A more speculative, but intensely interesting subject is the prevalence of life in the universe. We know of only one example—our own Earth. Whether the phenomenon of life is unique, rare, or a common occurrence appears now to be a problem that can be approached in a preliminary way. Instruments specifically designed for this purpose, with spatial resolution of the order of 10^{-3} to 10^{-4} arcsec, could obtain images of other solar systems. Furthermore, they could carry out spectrophotometric analyses of the chemical composition of the planetary atmospheres.

For our own Earth, the oxygen concentration is far higher than one would expect if it were near chemical equilibrium. Nearly two aeons ago, the oxygen concentration was enriched dramatically by the action of living organisms. A similarly high oxygen concentration, if found in the atmosphere of another planet, would be highly suggestive of the presence of biological activity there as well. Our knowledge of how planets form and how life arises is slight, and surprises would surely result from such a program of exploration.

A further, more specialized search involves intelligent life. Such a search has a different character than the broader-based search for life and is not addressed in this study. Obviously, if the NASA Search for Extra-terrestrial Intelligence (SETI) program were to find radio signals from another planetary system, it would be a tremendously exciting and significant event. In the case of the more restricted study described here, even the hint of life in another planetary system would trigger a new era in planetary research.

RELATIONSHIP TO OTHER DISCIPLINES OF SPACE SCIENCE

Astronomical instrumentation is becoming increasingly sophisticated and powerful in the high-precision measurement of objects far beyond our immediate solar system. This remote-sensing capability of astronomy can clearly be applied to other space science disciplines.

Sensitive cameras and spectrographs designed for high spatial and spectral resolution of cosmic phenomena allow imaging and spectroscopy of solar system objects that contribute to studies of the chemistry and physics of planetary atmospheres and magnetospheres. Images and spectroscopy of planets, circumstellar disks, and bipolar outflows from young objects can help to unravel the sequence of events leading to formation of stars and planetary systems. Studies of the magnetic activity of stars, and even direct imaging of stellar surfaces, enable us to extrapolate physical theories of solar activity and magnetic dynamos to objects like our Earth. Eventually, these studies may contribute to a deeper understanding of climatic variations on Earth, of the terrestrial dynamo action (including periodic reversals of the Earth's magnetic field), and of other phenomena of long-term importance to life on Earth.

3
Astronomy and Astrophysics in 1995: Expected Status

OVERVIEW

Astronomy has advanced dramatically over the past two to three decades, stimulated by the development of major ground-based facilities and the introduction of space techniques. A prominent feature of the astronomy program has been its reliance on detailed, long-term planning, exemplified most recently by the Astronomy Survey Committee report, *Astronomy and Astrophysics for the 1980s*. Looking forward to 1995, the task group anticipates progress in realizing the plans detailed in that report. Such progress will involve the principal elements of the Shuttle-based program, as well as HST, GRO, AXAF, and SIRTF, which form the first family of comprehensive observatories in space and have been called "the Great Observatories." These instruments will provide a powerful capability for detailed astrophysical studies at optical, gamma-ray, x-ray, and infrared wavelengths. Thus, the next decade will see space astronomy move strongly from an era of exploration to a program of in-depth study.

The Great Observatories are the foundation of this program. Their effective use will require the development of new state-of-the-art instruments for these telescopes during their useful lifetime and the replacement of the entire observatory when its useful life ends, due either to out-of-date components or scientific obsolescence. In the case of out-of-date components, it might be adequate simply to replace the observatory with one of equal capabilities. Replacement could also be seen as an opportunity for improvements that would make the observatory more capable to investigate the astrophysical questions of greatest interest at the time. Scientific obsolescence can occur either when the unique capabilities for which an observatory was built have been fully exploited or when general improvement of other observational capabilities diminishes the facility's productivity. The core program that the task group envisages is one in which we would attempt to maintain the four observatories as forefront research instruments. Different considerations will apply to the different wavelength regions.

Along with this evolution of technical means and scientific aims, the task group foresees two other trends—a further shift in emphasis from ground-based to space observations and a continued strengthening of the international base of space astronomy, with the realization of major European and Japanese missions, such as the Roentgen Satellite (ROSAT), the Infrared Space Observatory (ISO), and Japan's Explorer-class x-ray satellite (ASTRO-C).

In the event that the international base of space astronomy continues to strengthen, as present indications in Europe and Japan suggest, the task group also looks forward to further improvement in the cooperation and planning of the major programs of this era on a worldwide basis. Such a development would be fully in keeping with the international nature of astronomy.

During the coming decade, it will be vital to:

- maintain adequate support for the ongoing, long-lived observatories;
- provide, not only for major new initiatives, but also for medium- and small-scale missions, for supporting analysis, and for research and development; and
- introduce substantially less expensive ways of conducting space activities so that new facilities such as those outlined in Chapter 4 can become a reality.

RADIO ASTRONOMY

The results of extending observational capabilities into new domains are well illustrated by examining the recent history of radio astronomy. Naively, one might have expected that the relatively long wavelengths characteristic of the radio spectrum would have precluded the attainment of high angular resolution. This turned out not to be the case, because of the exquisite control of the time domain achieved by modern electronic techniques. Starting in the late 1960s, very long baseline interferometry (VLBI) extended the effective size of observing apertures to worldwide dimensions. Angular resolution of a milliarcsecond became commonplace, superior to the angular resolution achieved in any other part of the electromagnetic spectrum, and a number of new discoveries were made. The relativistic jet phenomenon was clarified and shown to have alignments over scale factors differing by more than a million. Two puzzling phenomena—velocities that apparently exceed the speed of light, and interstellar masers—still remain unexplained; both could potentially influence our basic assumptions about physical phenomena in the universe.

The first VLBI observations used simple two- and three-element configurations, but it rapidly became evident that more complete arrays were needed to give reliable pictures on the milliarcsecond scale. This led, eventually, to the establishment of the VLBI array project of the National Radio Astronomy Observatory (NRAO), which is expected to be in use by 1995. This gives the angular resolution of an aperture nearly equal to the Earth's diameter, provided a source is high in the northern sky. The problems at hand—reading down toward the site of acceleration of the relativistic jets, for example—require still longer baselines, but the Earth is too small. This led to the concept of space VLBI, and it is expected that the launching of QUASAT in mid-1995 will represent an important early step in exploring this new observational realm. The orbiting antenna of QUASAT will have an apogee of the order of 20,000 km, which will be further extended by earthbound telescopes covering another 10,000-km baseline. The wavelengths to be used are 1, 2, and 6 cm. Images will comprise up to 2000 × 2000 picture elements (pixels), with pixel dimensions as small as 90 μarcsec, and perhaps 40 μarcsec if coordinated international programs are effected. This will probably provide a first step toward understanding the problem of the "synchrotron-self-Compton effect" that theoretically limits the brightness of sources to 10^{12}K. It will lead to greater understanding of superluminal motions in quasars and of the structure of interstellar and circumstellar masers. When used in conjunction with the VLBA, the system will have millijansky sensitivity, allowing the study of a vast array of interesting objects.

INFRARED AND SUBMILLIMETER ASTRONOMY

In 1983 the Infrared Astronomical Satellite (IRAS), the first cryogenically cooled telescope to orbit in space, catalogued approximately 300,000 new infrared sources. Among its major discoveries were dust systems orbiting Vega and many other nearby stars; a "cirrus" cloud component on a galactic scale extending well above and below the galactic plane; two new zodiacal dust bands straddling the ecliptic plane; hosts of infrared sources, which, because of their very low luminosities, appear to be solar-type stars in their earliest stages of formation; high infrared fluxes from interacting galaxies; and large numbers of galaxies that are orders of magnitude brighter at infrared wavelengths than in the optical domain and are rivaled in their energy production only by quasars.

There are galaxies that at first sight might appear much like our own Milky Way—a spiral galaxy among billions of others. At infrared wavelengths, however, these galaxies have exhibited an enormous population of recently formed stars still shrouded by clouds of the gas and dust from which they formed. Massive star formation appears to have taken place there in a sudden outburst of activity. We will need to understand the origins and underlying mechanisms at work in these starburst galaxies.

IRAS discovered disks of finely structured material orbiting about several nearby stars. These disks may provide clues to the way in which planetary systems are formed around a star. SIRTF will be able to study a number of these disks to determine the spatial distribution and chemical composition of this circumstellar matter.

In order to understand the various physical processes that result in infrared emission under these diverse conditions, we need considerably higher spatial resolution coupled with high sensitivity. We also need to obtain spectral information about the chemical structure and physical conditions in many of these sources.

The task group expects that by 1995 a number of major space- and ground-based infrared facilities will be available for these studies, as well as for future observations that we cannot yet anticipate. These facilities are described briefly below:

- Cosmic Background Explorer (COBE) will have searched for deviations from the isotropic microwave background, examining the conditions when galaxies and stars first formed. It will also search across the entire infrared domain, mapping the large-scale, diffuse features of the galaxy.
- The Space Infrared Telescope Facility (SIRTF) and Infrared Space Observatory (ISO) will draw upon advanced detector and detector-array technology to provide imaging at sensitivities orders of magnitude higher than IRAS, together with greatly enhanced spectroscopic capability.
- A near-infrared imaging and spectrometry experiment will extend the Hubble Space Telescope's (HST's) advantages of increased angular resolution and reduced background from space into the short-wavelength infrared region of 1 to 3 μm. At the 3-μm wavelength there is a natural window that opens on the universe. In that region of the spectrum both the optically scattered zodiacal light and the infrared emission from zodiacal dust are particularly low, and our search deep into the universe will extend out to exceptionally great distances.
- A new generation of large, ground-based, optimized infrared telescopes, such as the 10-m Keck telescope, will provide large collecting area for spectroscopic observations and high angular resolution ($\sim$0.2 arcsec) imaging with high sensitivity to point sources.
- A number of 10- to 15-m-class submillimeter telescopes will be in operation at high mountain sites, providing limited access to the submillimeter regime through selected atmospheric windows. Observations in the far-infrared and submillimeter spectrum carried out from aircraft and balloons will continue to fill in the spectral coverage at lower angular resolution and sensitivity.
- High-resolution spectroscopic studies will be possible throughout the submillimeter range with the 8-m-class Far-Infrared Space Telescope (FIRST).

The major shortcomings in infrared observational capabilities in the mid-1990s will be the limited angular resolution from space observatories other than HST, the lack of a major submillimeter space observatory, and the absence of milliarcsecond interferometric capability.

ULTRAVIOLET AND OPTICAL WAVELENGTHS

By 1995 one or two ground-based telescopes of the 15-m class are expected to have been in operation for several years. In addition, a few 8-m-class ground-based telescopes should have been developed by the United States and Japan, and located at good observing sites in the northern and southern hemispheres. Despite major efforts to improve astronomical seeing, angular resolution from the ground will be limited to about 1 arcsec, with occasional glimpses of 0.3- to 0.5-arcsec image quality. The Hubble Space Telescope (HST) was planned to overcome these limitations by providing better than 0.1-arcsec resolution in the ultraviolet and optical wavelength regions.

By 1995 the HST will have been in operation for nearly a decade. If funded adequately, it will by then be equipped with second-generation instruments approaching detector quantum efficiencies of 80 percent over the spectral range 1200 Å to 2 μm. By 1995 the Extreme Ultraviolet Explorer (EUVE) and ROSAT will have surveyed the sky between 80 and 900 Å and the Far Ultraviolet Spectroscopic Explorer (FUSE/LYMAN) telescopes will perform detailed studies of objects between 900 and 1200 Å.

By the mid-1990s important astrophysical problems to be addressed with ground-based and space facilities will be the determination of:

- An accurate cosmological distance scale.
- Evolutionary and chemical states of normal and active galaxies to modest red shifts.
- Dynamics of disk and bulge regions in nearby galaxies and the distribution of associated dark matter.
- Physical and chemical properties of the interstellar and intergalactic media including deuterium in the local neighborhood.
- Physical and chemical states of degenerate stars and hot plasmas in stellar chromospheres and coronas.
- Activity and mass loss in every kind of star and stellar system.
- Initial search for protoplanetary disks and planets.

These studies will be aided by:

- Deep surveys with detection capabilities of point sources to large red shifts in small angular areas.

- Follow-up to sources discovered in new surveys (COBE, EUVE, and ROSAT).

Although point sources will be detectable to very large red shift limits with the HST, studies of galaxies will be limited by the rapid dimming of surface brightness for these highly red-shifted objects. Thus, for studies of galaxies, the HST will not be capable of exploiting the ultraviolet/optical region over substantial cosmic look-back times. Moreover, by 1995 there will be many fundamental astrophysical problems that will be limited by the angular resolution capabilities even of the HST.

Undoubtedly the next step will require higher angular resolution combined with large collecting areas so that our studies can answer the many questions that will have arisen about the spatial structure of the objects that HST will likely discover and examine.

X-RAY ASTRONOMY

X-ray astronomy will achieve a substantial increase in observational capability with the anticipated launch of the Advanced X-ray Astrophysics Facility (AXAF) in about 1994. AXAF will be the first major observatory for x-ray astronomy to be operated on a long-term basis. It will permit detailed, high-angular-resolution (about 0.5 arcsec) studies at x-ray energies from about 0.1 to 10 keV. It will observe the broad range of astronomical objects shown by the Einstein Observatory to be x-ray emitters—a range that runs the gamut from coronas around nearby cool stars to distant quasars. AXAF will not only permit detection of distant Quasi-Steller Objects (QSOs) and galaxy clusters but also will enable moderate-resolution spectral analyses of those sources. These observations will begin to address problems such as the formation and evolution of galaxies and galaxy clusters.

In studies of brighter objects, AXAF will allow high-resolution x-ray spectra to be obtained for the first time on objects as diverse as supernova remnants, accreting neutron stars, black holes, and normal stars. Once again, the exploratory spectroscopic studies carried out with the Einstein Observatory herald the rich potential for discovery and detailed astrophysical measurements that AXAF will conduct. In detailed spectroscopy of bright halos of galaxies such as M87, AXAF would determine the composition, temperature, and density profiles of the halo; that would greatly constrain the origin and mass profile of the galaxy halo. It would then be possible to determine whether the halo and its apparently large, dark mass component was supplied by accretion from the surrounding cluster or was intrinsic to the galaxy. Similarly, studies of dark matter in other nearby galaxies and clusters of galaxies will be initiated by AXAF. More-sensitive instruments, however, will be required to extend these studies to more distant objects of cosmological interest. The overall sensitivity of AXAF is well matched to that of the major optical (HST), infrared (SIRTF), and radio (VLA and VLBA) observatory capabilities anticipated for the early 1990s. AXAF will draw heavily on observations made with these facilities as well as on the moderate or Explorer-class missions in x-ray astronomy launched in the period from 1987 to 1994. These include the following:

- *ROSAT* (1987): A West German imaging soft x-ray (~ 2 keV) telescope, and a British XUV telescope make up this mission. ROSAT will carry out the first all-sky survey in these wavebands at high sensitivity, as well as a program of pointed observations in which U.S. observers will be directly involved. This mission should observe 100,000 x-ray sources.
- *X-Ray Explorers* (1987 to 1992): These are planned for the study of relatively bright compact x-ray sources. In 1987, Japan will launch a large-area (about 0.5 m^2), nonimaging x-ray detector for broadband spectrophotometry and the study of rapid variability in x-ray sources. It will probe the physics of accreting neutron stars, black holes, and white dwarfs in our galaxy and beyond. The U.S. X-ray Timing Explorer (XTE), to be launched in 1992, will extend the high time-resolution studies of such compact galactic x-ray sources and permit similar studies to still higher photon energies. The internal structure of neutron stars will be probed with sustained observations of x-ray pulsars and bursters, and physical conditions in accretion disks will be explored.
- *Space Shuttle Experiments* (1988 to 1990): These will be carried out, primarily on the OSS-2 mission, to develop new concepts for large-area x-ray telescopes with greatly increased sensitivity. These short-duration flights with limited capability will be vital for the development of future x-ray missions.
- *XMM* (1995): This is a first-generation, high-throughput x-ray telescope now under study by the European Space Agency (ESA) as a "cornerstone" mission. It will have a large collecting

area (about 1 m^2) for imaging and spectroscopy of soft x-ray sources at moderate spatial and spectral resolution. XMM is thus strongly complementary to AXAF and will also serve as an ideal springboard for the planning of x-ray facilities of the future (see Chapter 4).

The powerful capabilities of AXAF and the wealth of fundamental problems it can address suggest that this facility will advance research for a long time to come. When the need arises for major refurbishment or when the technology of producing significantly larger diameter and still higher-resolution x-ray mirrors has sufficiently advanced, the entire AXAF facility could be upgraded or replaced.

GAMMA-RAY ASTRONOMY

Small Astronomy Satellite-2 (SAS-2) and COS-B, following earlier pioneering measurements, have provided maps of the high-energy gamma-ray emission from the galactic plane and an interpretation of the origin and propagation of the cosmic rays. In addition, they have made observations of pulsars as well as other galactic objects, measured the basic properties of extragalactic diffuse gamma radiation, and discovered gamma-ray emission from a quasar. Active galactic nuclei, the galactic center region, and various sources, including transients, have been observed in the low-energy gamma-ray region by telescopes on the High Energy Astronomical Observatory-3 (HEAO-3), the Solar Maximum Mission (SMM), and other instruments. Reported observations of gamma rays at energies above 10^{11} eV using ground-based techniques bear on the physics of pulsars and active galactic nuclei. Thus, over the entire spectral range from 10^5 to above 10^{14} eV, there are now interesting, known sources of gamma radiation.

The next major advance in gamma-ray astronomy will come with the launch of the Gamma Ray Observatory (GRO), scheduled for 1988. Covering the energy range from about 10^5 to 3×10^{10} eV, it will have a sensitivity at least an order of magnitude greater than previous experiments over this whole energy range and a viewing program designed to scan the entire sky. Currently, the lifetime of the GRO will be from 2 to 4 years; the lifetime would be extended if refueling capabilities are provided. With GRO, gamma-ray astronomers will obtain the first comprehensive full-sky survey in moderate detail. In particular, GRO should further open the low-energy gamma-ray astronomy window by recording nuclear emission lines and investigating the properties of gamma-ray bursts in more detail. At the same time, it will extend studies in high-energy gamma-ray astronomy to a far greater depth.

Results from GRO will undoubtedly help to focus requirements and objectives for future experiments, but the general thrust of gamma-ray astrophysics in the post-GRO period is already reasonably clear. The Astronomy Survey Committee in its report *Astronomy and Astrophysics for the 1980s* already anticipated the need for at least two gamma-ray instruments beyond those on GRO. An advanced high-energy gamma-ray telescope of very large area, high sensitivity, and high angular resolution will be needed for long-term observations of selected sources and regions of special interest. This will be necessary to achieve the statistical accuracy in the counting of gamma-ray photons required to resolve spatial and spectral features of sources and to analyze their variations. The field of view of the telescope need not be wide, and an appropriate goal for angular resolution is of the order of 1 to 2 arcmin. A high-resolution nuclear gamma-ray spectrometer would also be needed for the study of the gamma-ray lines from radioactivity in supernova remnants, positron annihilation in the galactic disk and in extragalactic sources, nuclear excitations caused by cosmic rays in dense matter, and nucleosyntheses in extragalactic supernovae. The instruments for this era are discussed in Chapter 4. It is envisioned that these instruments would be well into their development stage by 1995 and ready for a flight opportunity soon thereafter.

COSMIC-RAY ASTROPHYSICS

Cosmic-ray particles provide us with the only direct sampling of matter of extra-solar-system origin. The great range of particle energy, covering about 15 decades, and the variety of species, encompassing the nuclei of all elements, as well as electrons, positrons, and antiprotons, require a number of different observational approaches. These approaches must be pursued on the ground, above the atmosphere, and in deep space.

Most of the instruments flown previously on spacecraft are of relatively small size and are therefore restricted to studies at low energies (~ 1 GeV), where solar modulation strongly affects the particles' energy spectra. Several of these spacecraft are expected to remain active into the 1990s. These investigations are providing: (1) definitive measurements with very high precision of the elemental composition of galactic cosmic rays (up to iron) at low energies; (2) exploratory data on the isotopic abundances; (3) direct sampling of particles in interplanetary space that are accelerated in solar flares or in planetary magnetospheres; and (4) perhaps, if our current interpretation is correct, a sampling of local interstellar neutral gas through the observation of "anomalous" cosmic rays (i.e., helium, nitrogen, oxygen, and neon, and possibly other elements with high ionization potential that may be accelerated at the boundary of the heliosphere).

The continuation of these studies through the 1990s will be essential in order to understand the spatial and time variations in the cosmic-ray flux through changing levels of solar activity. Instruments of modestly increased size will be used during this period in order to enhance the accuracy of the observations and to explore the solar system outside the ecliptic plane. Most important will be studies of the nucleosynthetic origin of the low-energy cosmic rays through a more precise determination of their isotopic abundances.

A continuing series of observations of galactic cosmic rays at higher energies is being conducted on balloons and on the Space Shuttle. These provide important exploratory data, but definitive measurements have to await the availability of spacecraft capable of carrying very large payloads for extended periods of time (years). Objectives of such measurements are to determine: the isotopic abundances at high energies; the composition of the rare ultraheavy nuclei; the energy spectra of nuclei and electrons over a wide range; and the abundances of positrons and antiprotons. The Astronomy Survey Committee had recommended carrying out a series of such observations (*Astronomy and Astrophysics for the 1980s*), but only a few investigations are being implemented. Thus, by 1995 we will probably still be short of decisive data that could specify the character of particle acceleration in the galaxy, or that could define the nucleosynthesis processes leading to cosmic-ray matter and governing evolution of the galaxy. We will probably also lack definitive data that could further our understanding of the structure and properties of the interstellar medium, and that could address some fundamental cosmological questions such as matter/antimatter symmetry in the present universe.

The investigations listed in the following subsections are under way at present or are planned by 1995.

Interplanetary Spacecraft

Particle detectors on Pioneer 10 and 11 and Voyager 1 and 2 will continue to explore the solar system at ever-increasing distances from the Sun. Space scientists will use instruments on Interplanetary Monitoring Platform-8 (IMP-8) and perhaps on the Wind spacecraft of the International Solar-Terrestrial Program (ISTP/Wind) as references closer to Earth. The Ulysses Mission will provide in situ measurements outside the ecliptic plane for the first time. An important time period for cosmic-ray observations at low energies will be the next minimum in solar activity, around 1988-1989.

Cosmic-Ray Composition Explorer

The exposure of a sensitive particle detector onboard an Explorer-class mission outside the Earth's magnetosphere has been highly endorsed by the scientific community, and should be implemented before 1995. In this instrument a large-area solid-state detector telescope, combined with trajectory measuring devices, will obtain an order-of-magnitude improvement in sensitivity and resolution over previous instrumentation. We thus expect qualitatively new insight into the elemental and isotopic composition of a variety of particle populations at low energies ($\sim$1 GeV): galactic cosmic rays, the "anomalous" component, and nuclei accelerated in solar flares. The first two are thought to be rather contemporaneous samples of the interstellar medium, while the third is a more ancient sample that has been stored in the Sun for almost 5 billion years. The comparison between the compositions of these particle populations will therefore either reveal characteristics of the chemical evolution of the interstellar matter—for instance, a continuing metal enrichment—or indicate if the solar system composition is atypical, perhaps owing to the admixture of fresh supernova ejecta during its formation.

Space Shuttle and Space Station

A very large detector to measure elemental abundances and energy spectra of cosmic-ray nuclei at very high energies has been developed for sortie flights on the Space Shuttle. A first flight was performed in 1985, and several reflights are expected that will yield data on the cosmic-ray composition up to energies around 10^{13} to 10^{14} eV. These measurements will constrain models of galactic particle acceleration by determining, for instance, the energy dependence of the relative abundances of primary cosmic rays (e.g., the iron/carbon ratio). They will also provide decisive information on the interaction of particles and interstellar matter and fields by measuring the relative abundances of spallation-produced particles (e.g., light nuclei such as lithium, beryllium, and boron) over an extended energy range. A long- duration flight (approximately 1 year) of this same instrument could cover another decade in energy, close to the "knee" of the cosmic-ray spectrum around 10^{15} eV. This may be accomplished onboard a Space Station or Space Platform in its initial operational configuration. This experiment would reach into an energy regime where the cosmic-ray composition is entirely unexplored at present.

Instrument Development

A cosmic-ray facility centered around a large superconducting magnetic spectrometer in space has been proposed by the scientific community and is currently under study as a high-priority new initiative. The spectrometer will be the common, central component for a succession of specific investigations, directed, for instance, toward measurements of protons and electrons and their antiparticles, antiprotons, and positrons. Other investigations might include searches for heavy antimatter, and studies of isotopic composition at high energies. During the coming years, detector development for such a facility is expected to be in active progress. The first elements of the facility are to be deployed during the initial stage of the Space Station in the early 1990s.

GRAVITATIONAL PHYSICS

Gravity is one of the fundamental forces of nature. Its actions in the neighborhood of relatively small masses like the Sun and Earth are well approximated by Newton's laws. These laws, however, have been superseded by Einstein's general relativistic description of gravity. General relativity predicts small deviations from Newtonian behavior in weak gravitational fields. Some of these deviations are being measured accurately in the solar system. General relativity also predicts both new classes of phenomena in strong gravitational fields, and the generation of gravitational waves by accelerating massive objects such as compact stars in close binary orbits (e.g., the millisecond pulsar) and supernova explosive events in which enormous masses are hurled into space in gigantic explosions.*

One test of the first of these two classes of effects will be carried out in NASA's gravity probe B (GPB). This experiment in earth orbit will involve a rapidly rotating, superconducting, niobium-coated sphere whose pole direction will be affected a minuscule amount—a precession of 0.04 arcsec/yr—by the rotation of the Earth. The detection of this change should provide an important test for the magnetic-like effects of mass predicted by general relativity.

The already successful solar system tests of general relativity could be extended to second (post-Newtonian) order in the solar potential. Instruments to perform such tests could be developed by the mid-1990s. NASA has also considered a mission (STARPROBE) to place a precision clock in a near-Sun orbit to measure the second-order gravitational red shift. In addition, a proposal exists to build a small astrometric optical interferometer (POINTS) to measure the deflection of starlight by the Sun. Each of these experiments also has many other scientific objectives. Should either experiment be conducted and fail to confirm the predictions of general relativity, we would be forced to rethink our understanding of a fundamental part of physics and its implications for astrophysics.

Tests for the existence of at least a restricted frequency range of gravitational waves will be provided by spacecraft equipped with dual-frequency transponders in distant trajectories throughout the solar system. Passage of a gravitational wave between these spacecraft and Earth should provide a signature in the arrival time of transponder response uniquely identifying the passage of such a wave. Current tests are orders of magnitude too insensitive to

---

* A separate task group on fundamental physics and chemistry has addressed the topic of gravitational physics in the context of this study. Its findings are published in the volume *Space Science in the Twenty-First Century: Fundamental Physics and Chemistry*. However, this subject is of great interest to astronomers and astrophysicists and is therefore discussed briefly here.

detect such waves, but future observations of this kind may be sufficiently improved to provide a useful complement of ground-based instruments developed to detect gravitational waves of far higher frequencies. To date, the only indication that we have for the existence of gravitational waves comes from the secular decrease in the orbital period of the binary radio pulsar.

NASA OPERATIONAL STATUS

The post-1995 program for astrophysics is strongly dependent on the state of NASA technologies. The task group anticipates routine Shuttle operation and the ability to transport large masses to optimum scientific orbits.

With regard to the state of astrophysics projects, the task group assumes the following:

- HST will be operational with at least one, and probably two, instrument upgrades completed.
- GRO will be nearing the end of its operating lifetime with about 50 percent of the scientific instrumentation operating.
- AXAF will be in orbit and nearing the time for its first servicing.
- SIRTF will be operating with the first cryogen servicing planned for the near term.
- LDR will be undergoing intensive design study.
- Among the moderate-scale projects, FUSE, QUASAT, and HTM will also be operating.

INTERNATIONAL PROGRAMS

There are a number of programs originating outside the United States in Europe, Japan, and the Soviet Union. These should be listed here as well. Some have already been approved; others are planned or only under discussion. Further, we must understand that some of the missions both within and outside the United States will be launched before 1995, while some might be delayed. Others might be logical continuations of missions started before 1995. A clear-cut demarcation therefore cannot be drawn at 1995. A further complication is the difference between international European missions as distinct from autonomous national European missions. Lumping all these varied efforts together, we then have the following array of instruments:

- *Radio*: There is a Japanese version of QUASAT (space VLBI) under study by a working group. The plan is to launch a 5- to 10-m diameter antenna at 20 GHz into a 500- to 40,000-km orbit around the year 1995. There also is an approved Soviet plan, RADIOASTRON, to launch several radio telescopes into orbit for VLBI purposes, possibly coordinated with QUASAT.
- *Infrared and Submillimeter*: ESA is constructing an Infrared Space Observatory (ISO) to be launched around 1992. It will have the capability for producing maps and obtaining low-resolution ($R = 1000$) spectra over a wavelength range of 2 to 200 μm. There will also be polarization sensing capabilities aboard. A further effort is the Far Infrared Space Telescope (FIRST), a heterodyne spectroscopy mission planned for the mid-1990s. It will have high-precision capability for determining chemical composition and velocity structure within cool clouds, at submillimeter wavelengths.
- *Optical Astronomy*: The ESA astrometric mission HIPPARCOS will improve optical astrometric accuracy by an order of magnitude and survey about 100,000 stars. It will provide parallaxes for a variety of astrophysical studies including a refined color-luminosity relationship and an improved distance scale based on direct measurements of a few Cepheid and RR Lyrae variables. HIPPARCOS will also provide proper motions with an uncertainty of about 1.5 milliarcsec/yr. These will find use in the study of the dynamics of the Hyades and in the determination of the birthplaces of young stars. In Japan, a survey-type ultraviolet telescope, UVSAT, is on the menu of the series of small-to-moderate satellite missions. The hope is that UVSAT will complement HST with a sub-Lyman-alpha, moderate spatial- and spectral-resolution capability. Currently, it is intended for launch in 1995; the exact schedule will depend on that of Japan's next generation of ground-based optical telescopes.
- *High-Energy Astrophysics*: Japan's Explorer-class x-ray satellite, ASTRO-C, is being prepared for launch in 1987 in a collaborative venture with the United Kingdom. Whenever possible, Japanese satellites will carry a small gamma-ray burst monitor as well, as a matter of policy. The gamma burst monitor for ASTRO-C will be provided by Los Alamos National Laboratory. ASTRO-C will be nonimaging with an area of 0.5 m^2. ASTRO-C

may be followed by ASTRO-D. The plan is to carry an imaging device with an area in excess of 500 cm^2 and with moderate angular resolution (about 1 arcmin).

The West German ROSAT mission of 1987 will be followed by a hard x-ray mission (SAX) being prepared in Italy and by a gamma-ray imaging facility, SIGMA, being prepared in France for flight on a Soviet mission.

In addition, ESA is planning a high-throughput x-ray spectrometer, XMM, within their future space science plan (Horizon 2000 Program).

4
New Initiatives

THE WAY FORWARD

The previous chapters have outlined how astronomers have developed a totally new view of the universe and have projected the expected state of space astronomy in 1995. Our observational capabilities have increased steadily. New phenomena have been revealed at each advance in sensitivity, spectrum coverage, and angular resolution. Increasingly, the complementarity of observations in different parts of the spectrum has been revealed, emphasizing the view that access across the electromagnetic spectrum is essential in advancing our knowledge of the universe. The Great Observatory program, involving the HST, SIRTF, AXAF, and GRO by the 1990s, meets many of the present needs. The task group has assumed that this core program, the culmination of many years of planning and experimentation, will have been implemented by 1995.

The program of new initiatives for the era 1995 to 2015 focuses on improvements in capabilities in two areas: higher angular resolution and greater collecting area.

The first of these, high-resolution imaging, requires development of interferometric arrays to synthesize large apertures. The goal varies with wavelength, but in general the aim is to work toward microarcsecond imaging at radio and optical wavelengths (including the ultraviolet) and to obtain milliarcsecond resolution or better at infrared wavelengths. This would mark the logical continuation of the program started by Galileo, when he launched the modern era of astronomy. Galileo's telescope, which showed details that were a factor of 10 finer than the human eye could see, started a process that still goes on today. As knowledge at one level of detail is consolidated, the new questions this knowledge raises justify further explorations. The program for high-resolution imaging, described in the following section, represents the latest step in this evolution.

The second general need, for greater collecting area, is more accurately described as the need for high-throughput instruments. The techniques to attain this vary greatly from one part of the electromagnetic spectrum to another, but philosophically these, too, are a continuation of Galileo's program in a different aspect. The greater collecting area of Galileo's telescope again allowed the observation of fainter objects than the eye could see, and that too has continued as a major thrust in astronomy through the building of larger telescopes and better detectors. The needs for high-throughput instruments can be identified at submillimeter wavelengths, in the optical, ultraviolet, and x-ray domains, extending down to gamma rays and including the particle detectors of the cosmic-ray astronomers. These needs are detailed in a separate section. The instruments described there involve new technologies in some cases, but are not beyond the projected capabilities for 1995 and thereafter. In some instances deployment of the instruments could be facilitated by partial fabrication and assembly in space. The instruments are described in order of decreasing wavelength, but no priority ordering is implied.

The initiatives described here will demand new technology and new capabilities in space. The only way to make sharper diffraction-limited images will be to lengthen the baseline over which the wave front is sampled. This can be achieved directly by increasing the diameter of the telescope reflector, or indirectly by coupling together radiation or signals from widely spaced reflectors. To overcome quantum noise we need to provide telescopes with larger collecting areas. These considerations set the direction for future evolution toward still larger telescopes, and the period beyond 1995 will bring new opportunities to pursue this evolution.

Adaptations planned for the Space Transportation System

(STS) will allow the launching of telescopes and components considerably larger than the limit set by the Shuttle cargo bay. The Space Station, or ultimately a lunar base, will offer an arena in which very large telescopes or arrays can be assembled, tested, and fine-tuned in their operating environment. Transfer vehicles will allow these telescopes to be placed and serviced in optimum orbits.

HIGH-RESOLUTION INTERFEROMETRY

Introduction

When Michelson invented the stellar interferometer in 1920, the promise of the technique was clear, and one of its principal technical obstacles was equally clear. The sizes of stars would be estimated from their temperature and magnitude, and the interferometer could, in principle, provide the necessary milliarcsecond (or better) angular reduction; unfortunately, astronomy had to be carried out at the bottom of the Earth's atmosphere, whose turbulent behavior so perturbed the incoming wave fronts that phase coherence was lost over apertures larger than a few centimeters and for times longer than a few milliseconds. The method was little used until the radio astronomers were able to adapt and refine the technique. Very long baseline radio interferometry (VLBI) is now used routinely, for example, to achieve astrometric accuracy of 10 to 100 μarcsec (depending on source separation and structure). The Very Large Array (VLA) in Socorro, New Mexico, represents the present culmination of radio inferometry in the form known as aperture synthesis, in which complete Fourier information is obtained over an aperture 35 km in diameter, and images are obtained with correspondingly high resolution (0.1 arcsec at 2-cm wavelength). The VLBA, now under construction, will extend imaging capabilities to better than a milliarcsecond by the same aperture-synthesis methods, but with an aperture more than one hundredfold larger. The QUASAT mission will extend the aperture size still further.

The advent of space science now raises the promise of using the clarity of space to achieve similar capabilities in the infrared, visual, and ultraviolet regions of the spectrum. Except for technical details, none of them fundamental, the concept of aperture synthesis carries over to the shorter wavelengths of the optical domain. The ultimate goal is to achieve images with 1-μarcsec resolution, but this millionfold increase from our current resolution on Earth is probably not achievable in a single step. A thousandfold increase, however—comparable to the step from the VLA to the VLBA—does seem to be a reasonable first step. Progress toward still higher angular resolution might then come with the establishment of this technology. Much of the necessary technology for imaging and astrometric interferometry is held in common. Astrometric developments with the same or a similar instrument would probably allow microarcsecond accuracy. An astrometric instrument with microarcsecond accuracy would have numerous applications including a light-deflection test of general relativity sensitive to the effect of the square of the solar potential. Such a test would be the first "second-order" solar system test of general relativity. Other possible scientific uses include: a search for extra-solar planetary systems; a direct determination of the Cepheid distance scale; the determination of the masses of stars in binary systems and those close enough to apply the method of perspective acceleration; parallax measurements yielding both absolute stellar magnitudes and, in conjunction with mass estimates and other data, a sharpened mass-color-luminosity relation; a study of mass distribution in the galaxy (and thus an improved understanding of its dark-matter content); a strictly geometric (i.e., coordinate and parallax) determination of the membership of star clusters (particularly useful in the case of peculiar stars such as blue stragglers and Wolf-Rayet stars); and a bound on or measurement of quasar relative motions.

    A workshop held in Cambridge, Massachusetts, in October 1985 reviewed the prospects for interferometry and concluded that "imaging interferometry in space will ultimately play a central role in astrophysics, a role comparable in significance to that played by space observations at x-ray and infrared wavelengths." The current level of effort in space interferometry is extremely small, and the workshop concluded that an orderly program had to be constructed. This would have to include the following: technological development of structures, spacecraft control, and optical technology; the study of a variety of instrumental concepts; the flight of small interferometers; and the formulation of a long-range program leading to a major observatory-class instrument.

    The earliest observation in the milliarcsecond range would be intensely interesting, and as one progresses to the microarcsecond level further dramatic results can be anticipated. Figure 4.1 shows some of the phenomena that can be studied at various scale sizes as a function of distance. Practically every class of object of astrophysical interest appears in the diagram. Constant angular resolution in this representation follows the diagonal lines shown, and one can see that the HST reaches only some of the region of interest. A milliarcsecond instrument reaches the resolution range for several major classes of object, including stars, novae, star-forming regions, and the broad-line region of quasars and active galactic nuclei (AGN). When one progresses beyond, toward microarcsecond resolution, the fields of interest become progressively richer.

The quasar/active galactic nucleus problem can be used as an illustrative example. These are probably related phenomena, differing only in scale: our own galaxy has a moderately active, compact nucleus, possibly containing a black hole of perhaps a million solar masses. Seyfert galaxies have more active nuclei, while quasars with their spectacularly high energy output are the most active of all. The power output from these objects is derived from gravitational energy as matter in the surrounding galaxy falls into the central black hole. The matter cannot be pulled in directly since it has angular momentum, so it settles into an accretion disk, where the angular momentum can be transported outward as the material spirals inward. The phenomena are coupled and lie at the forefront of our physical knowledge. One of the unexpected consequences is the generation of the highly collimated jets observed by the radio astronomers. These clearly involve the acceleration of bulk matter to relativistic velocities. The optical and x-ray fluctuations of quasar brightness hint at the existence of other dynamical phenomena that can best be studied by viewing the phenomena directly.

The angular resolution required to study the structures in an active galactic nucleus is illustrated in Figure 4.2. For M87 (Virgo 4), the closest highly active galactic nucleus, a milliarcsecond instrument would reach close to the accretion disk; if a resolution of even a few tens of microarcseconds could be achieved, the actual details of the accretion disk phenomena could be studied. The dimensions characteristic of the black hole itself are still beyond the observable horizon until we can achieve resolution somewhat better than a microarcsecond. Whether the central singularity is

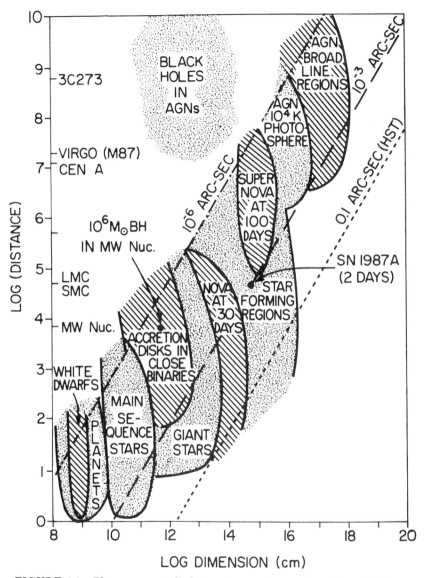

FIGURE 4.1 Phenomena studied at various scales as a function of distance.

actually a black hole or is a still more exotic form of matter cannot be said, but it is clear that in studying this class of phenomena we would be drawn into a new domain of physics: the regime of strong-field gravitation.

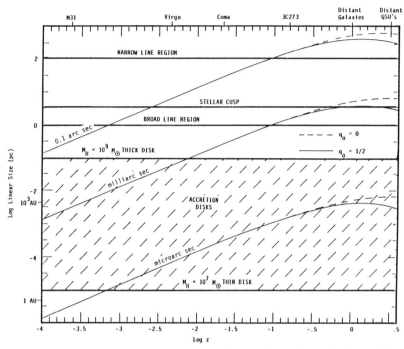

FIGURE 4.2 Angular resolution required to study active galactic nuclei.

Optical and Infrared Interferometric Arrays

The infrared and optical domains of the electromagnetic spectrum (including ultraviolet wavelengths as part of the optical spectrum) are ripe for investigation using the high-angular-resolution capabilities of aperture-synthesis interferometry. The manifold possibilities have been illustrated in Figure 4.1, but a program must be formulated that can make these a reality. Both ground-based and space-based interferometry need more intensive development. Clearly, instruments should be located on the ground when that is feasible, as it probably is in parts of the infrared spectrum. Ultraviolet interferometry, however, can never be done from the ground, nor can interferometry in those parts of the infrared spectrum that are blocked by the Earth's atmosphere. Likewise, throughout the visual part of the spectrum the disturbing influence of the Earth's atmosphere is so great that milliarcsecond and microarcsecond resolution seems most unlikely. Thus, visual interferometry will also depend upon space-based instruments.

The first major instrument in space might well be one composed of a number of medium-sized telescopes, mounted on a compensating structure that would maintain their phase coherence. The exact configuration of the array, the number and size of its elements, and the array location (low earth orbit, geosynchronous, L5, or the Moon) remain to be defined, but a reasonable representation is illustrated schematically in Figure 4.3. This shows nine elements on a tetrahedral truss, with the image-processing equipment at the fourth vertex. The mass distribution would be chosen to yield a zero quadrupole moment, thus reducing gravity-gradient torques. The dimensions might well be in the 50- to 100-m range, and the elements might be 1.5 m in diameter. Such an instrument, if 100 m in dimension, would give angular resolution of 1 milliarcsec at 5000 Å; at 2000 Å, the resolution would be 0.4 milliarcsec. The collecting area would equal that of a 4.5-m-diameter telescope, so its sensitivity would be comparable with that of any ground-based telescope operating today.

An entirely different concept might consist of free-flying telescopes, whose connection is only by laser beam. This might be called a Long-Baseline Optical Space Interferometer (LBOSI) and would achieve exceedingly high angular resolution on both bright and faint astronomical objects over a wide wavelength region (perhaps as much as 0.2 to 500 μm; ultraviolet to submillimetric). The LBOSI would consist of two or more large (8-m-class) diffraction-limited telescopes separated by variable baselines. For a maximum separation of 100 km, an angular resolution of 1 μarcsec would be achieved. There are major technical hurdles to be overcome, however. The most serious one is precision station keeping and attitude control. Studies of the limits of these technologies should be undertaken. This concept of free-flying telescopes is probably more difficult and more expensive than the monolithic tetrahedral concept shown in Figure 4.3, but if it is technically feasible it would be an exciting instrument to use. Expandability to even longer baselines is a definite advantage of this type of array.

Development of even these first-generation optical interferometers would require extensive technical progress. This would include the development of means to measure rapidly and accurately the positions of the elements, and to compensate for path length changes in a dynamically stable fashion. Methods for detecting fringes simultaneously at many wavelengths in the presence of photon and background noise, with an adequate field

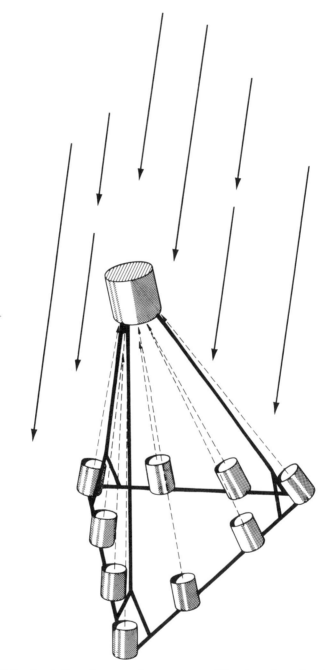

FIGURE 4.3 A (large) space telescope array: Nine 1- to 2-m-class telescopes on a 50- to 100-m tetrahedral truss.

of view, must also be developed. Since true images are desired, the precision of the path lengths would have to be kept to a small fraction of a wavelength during an integration period. This may be done by using unresolved stars in the wide field of view available to these space interferometers for sensing phasing errors. Optical and infrared interferometry may profit from the extensive phase closure and self-calibration techniques developed for radio aperture-synthesis instruments, and their adaptation must be carefully worked out.

Given the extent of this technical challenge, the task group recommends outlining a developmental program over the next decade. Some of this developmental work can be carried out on the ground, but some will require space experiments. These would certainly include small interferometers, probably having only two elements and capable of assembly in space.

Extensions of Orbiting Radio VLBI

In the model for the space science status in 1995, summarized in Chapter 3, the radio astronomy mission QUASAT was projected, extending the VLBI imaging capability to baselines of about 3 earth diameters. At present, the Soviet Union is planning a series of VLBI satellites (RADIOASTRON), and they have indicated a willingness to coordinate their plans with NASA and ESA. Furthermore, Japan is studying the feasibility of launching a VLBI satellite. Again, with proper coordination, we can expect an augmentation of resolving power surpassing 10 μarcsec. Radio astronomy has had a history of uncovering surprises, and this extension of QUASAT may be expected to continue this tradition. If the results are provocative, an extended, more ambitious array might be contemplated for the era beyond 2000.

The natural limits of radio aperture synthesis are set by the interstellar medium (ISM), which is an inhomogeneous plasma. At the higher galactic latitudes, above the obscuration that affects optical and ultraviolet extinction of intragalactic objects, the seeing limits imposed by the ISM are not serious down to a resolution of a microarcsecond at a wavelength of 1 or 2 cm. This means that an array of radio telescopes can be envisaged, extending to baselines of 100 earth diameters or so, yielding an angular resolution of 2 μarcsec at 1.3-cm wavelength. If the expected advance of microwave electronics proceeds, the minimum working wavelength might well be 3 mm and the maximum angular resolution would then be 0.5 μarcsec. The scientific possibilities with respect to the study of active galactic nuclei can be seen by referring to Figures 4.2 and 4.3.

The technical realization of plans to extend radio VLBI will depend on the development of space technology. A current study, known as ASTROARRAY, projects a total of 6 to 12 antennas in a variety of orbits chosen to give full aperture synthesis. The size of the individual antennas might well be of the order of 30 m, a size consistent with the projected state of the art in the year 2000. VLBI is currently a strongly international activity, and ASTROARRAY, if it becomes a reality, would almost certainly be international in character.

Future Developments

The direction of developments in interferometry can be unexpected, as the sudden advent of VLBI demonstrated. The viewing of the world in Fourier transform space has been a strong current in much of modern science, and over the next 30 years one may well see new possibilities open up as technology advances. Microarcsecond resolution at x-ray wavelengths, for example, would be enormously exciting, and even though x-ray interferometry is a technically difficult concept, there appear to be no fundamental physical barriers. An x-ray interferometer with microarcsecond resolution would be of the order of a few tens of meters in diameter. Although such an instrument has barely been conceptualized yet, a prudent program would keep an awareness of its potential. Flexibility of reaction is essential within the space science program in this respect, especially since relevant experiments on a much smaller scale could be carried out.

HIGH-THROUGHPUT INSTRUMENTS

The Large Deployable Reflector (LDR)

The Large Deployable Reflector (LDR) will be a 20- to 30-m-aperture telescope dedicated to far-infrared and submillimeter observations from space. Assembled in space, it will operate between about 30 and 1000 μm (1 mm). It will provide angular resolution of 1 to 2 arcsec at 100 μm and 0.3 to 0.6 arcsec at 30 μm, i.e., comparable with that currently achieved in the optical region of the spectrum. It would be engineered to be capable of spectral resolution on the order of 3×10^6.

The LDR is ideally suited to the study of the following:

- Collapse of protostellar clouds to form stars.
- Formation of protoplanetary clouds.
- Transition in the mass distribution within multiple-member systems from binary star systems to planetary systems.

A number of other areas where results are less predictable but potentially of fundamental significance include the following:

- Spectral and spatial deviations from an isotropic blackbody cosmic background spectrum.
- Kinematics and evolution of galaxies, clusters, and active galactic nuclei.
- Galactic nebulosity of all kinds.

The first far-infrared/submillimeter observations were carried out from balloons, from NASA's Lear Jet, and from Aerobee rockets in the late 1960s and early 1970s. These led to the recognition that galactic gas and dust complexes radiate the bulk of their energy at wavelengths around 100 μm and that the center of the galaxy radiates most of its energy at these wavelengths. It was also found that quasars and external galaxies, particularly Seyfert galaxies, are powerful emitters of far-infrared radiation, often emitting the bulk of their energy in the far infrared.

In the late 1970s more detailed investigations were started with the use of telescopes aboard NASA's Kuiper Airborne Observatory, aboard the Lear Jet, and on balloon flights carried out by groups from several countries. These observations resulted in maps of galactic sources at resolution down to 1 arcmin at 100 μm, and detailed spectra of wide classes of astronomical sources. These included regions of star formation, planetary nebulae, highly evolved stars, and external galaxies. With these spectra we gained an insight into physical conditions prevailing in these sources. The long wavelengths employed permitted penetration of thick dust clouds often associated with strong infrared/submillimeter sources and allowed us to study the physics of high-velocity shocks that might be responsible for the initial compression of gas presaging the collapse of a dense cloud. This collapse is thought ultimately to lead to a small cluster of newborn stars.

While highly informative, these studies have persistently suffered from the low angular resolution afforded by the small telescopes borne aloft by airplanes and balloons. The LDR would increase the angular resolution by a factor of 20 over that available in 1985. At 100-μm wavelengths, that would result in an angular resolution of the order of 1 to 2 arcsec. At 30 μm the resolution would be a factor of 3 better. The 20- to 30-m aperture of this monolithic telescope will also offer an increased light-gathering power of the order of 400 to 900 over and above the capabilities of the 1-m class telescopes in existence in 1985.

The LDR will be a natural sequel to the IRAS mission flown in 1983, the Cosmic Background Explorer (COBE) satellite to be launched in the late 1980s, and the Infrared Space Observatory (ISO) and SIRTF missions planned for launch in the early 1990s. IRAS discovered a large number of extragalactic sources whose emissions peak at or beyond 100 μm. Currently planned missions are not likely to have the sensitivity at several hundred microns to detect these sources, although SIRTF may be able to extend the wavelength coverage to 200 μm. The COBE mission will be able to determine the absolute background radiation across the entire infrared, submillimeter, and millimeter domain. It will not, however, be able to provide angular resolution in excess of a few arcminutes in the relevant millimeter domain where the microwave background radiation remaining from the initial cosmic explosion is to be found. LDR will be able to improve on that resolution by a factor of 30 or 40. These technical capabilities will lead to enormous scientific advances. A smaller mission with capabilities leading up to those envisaged for LDR is a Far Infrared Space Telescope (FIRST), currently contemplated by ESA.

An 8- to 16-m Telescope for Ultraviolet, Optical, and Infrared Wavelengths

A large-aperture space telescope for the ultraviolet, optical, and infrared regions has immense scientific potential. The need for such a telescope will be very high after 10 to 20 years of use of HST and ground-based 8- to 10-m-class telescopes. Even now we see that some of the most fundamental of all astronomical questions will require the power of a filled-aperture telescope of 8- to 16-m diameter designed to cover a wavelength range of 912 Å to 30 μm, with ambient cooling to 100K to maximize the infrared performance.

With the HST and SIRTF still to be launched, and the anticipated wealth of data not yet analyzed, it is difficult but not premature to formulate a detailed concept of such a large-scale telescope for the ultraviolet, optical, and infrared regions. This telescope will also complement a space interferometer. The diffraction-limited resolution of a 15-m telescope is 6 times sharper than that of the HST; it would be far more sensitive, both because of its greater collecting area and because of the small size of its diffraction-limited image. The image is 1/100 arcsec in diameter for a 15-m telescope in visible light. At this resolution the reflected zodical background sky limit is reached at magnitude 33, about 11 magnitudes fainter than ground-based telescopes and 4 magnitudes fainter than the HST.

The large telescope would not simply be a scaled-up HST. Its infrared performance would be optimized by cooling the optics to at least the lower limit of passive radiation methods, about 100K. Similarly, the optical performance would be greatly enhanced by a wide-field optical design suitable for large, high-resolution detector mosaics and multiple-object spectroscopy. Images with 10^{10} picture elements are possible with present optical designs, and the technology for charge-coupled device (CCD) mosaics this large is advancing.

A large range of scientific problems could be undertaken only by a telescope of this type. The combination of light-gathering power and resolution offered by such a telescope, equipped with advanced spectrographs and detectors, would lead to a quantum leap in our understanding of some of the most fundamental questions in astronomy.

Scientific Objectives for an 8- to 16-m Telescope

Galaxy Formation and Distant Quasars. Consider first the formation of galaxies and the earliest generation of stars, when the process of making new elements from primordial hydrogen and helium began. Light from stars and supernova outbursts from this period must be reaching us still, but it is too faint and too strongly red-shifted to detect from the ground. The actual red shift at which the bulk of galaxies formed is not known but is suspected to be in the 3 to 10 range, with strong evolution of galaxies known to be continuing to red shifts as small as 0.5. At red shifts of 3 to 10, we must search for primeval galaxies and their supernovae in the infrared at wavelengths of 1 to 5 μm. However, from the ground, atmospheric emission and absorption are prohibitive in this region. At 1.6 μm the "sky" is some 200 times darker in space. Even greater gains can be made by taking advantage of an extraordinarily dark window in the 2- to 4-μm range. The limiting magnitude from space in both the optical and infrared windows is set over the whole sky by a background of sunlight reflected or absorbed and reradiated by zodiacal particles. Between the reflection and emission peaks lies a window ideal for exploring the early universe.

In order to exploit this window, the telescope must be cooled so that its own thermal emission is negligible. It must also be large enough to detect very weak signals and to form sharp diffraction-limited images in the infrared. At 8- to 16-m aperture, its resolution at 3 m will be similar to the optical resolution of the HST. The distant primeval galaxies will then be resolved, and their structure can then be compared with that of nearby galaxies to understand how galaxies evolve.

Active Galaxies and Quasars. Twenty years of study of active galaxies and quasars have whetted our appetites for some real information and understanding of the nuclear powerhouse, its immediate environment, and its effect on the host galaxy. This is particularly true of quasars. The tantalizing detail seen with the 1- to 20-milliarcsec resolution from radio VLBI observations needs to be complemented by ultraviolet-optical-infrared studies. A telescope of 16-m diameter can directly resolve structures as small as 5 milliarcsec or, equivalently, less than 0.5 parsecs (pc) at Virgo in the near ultraviolet (0.3 to 0.4 μm). Such resolutions allow direct imaging and spectroscopic studies to the very edge, if not into, the broad-line region in active galaxies and in the nearest quasars. They will also elucidate the fine details of the narrow-line region, even to high red shifts. In fact, the structure of the host galaxy can be imaged and studied spectroscopically in the visible with resolutions better than 100 pc for all red shifts, even those greater than 1, limited only by the brightness of the source and the telescope's light-gathering capability.

Furthermore, this resolving power allows for direct observations of the superluminal clouds in sources such as the quasar

3C273 and the radio galaxy 3C120 on scales of 3 to 30 milliarcsec. It also allows observations of the one-sided VLBI jets, again on 3- to 10-milliarcsec scales in, for example, the radio galaxies NGC 6251 and 3C120. Following the movement or changes in the properties of the superluminal clouds over several years could help us to clarify the nature of these enigmatic objects. Since synchrotron decay times are several hundred times shorter for optical emitting electrons than for radio-emitting ones, optical continuum monitoring of jets with an 8- to 16-m space telescope would give unique information about how the particle acceleration processes evolve with the outward propagation of the jets. In addition, the diagnostic tools of spectroscopy could be used to record the effects of the jet propagation on the ambient medium.

The ability to image these small and complex sources unencumbered by the difficulties in the model-dependent interpretation of interferometric data (particularly speckle) is a great advantage of directly imaging with a filled aperture. These studies would complement and enhance programs using interferometers in space to study the highest surface brightness features associated with a source's central powerhouse.

Evolution of Galaxies. A large space telescope will have unique capabilities for studying the structure of galaxies at intermediate red shifts of 0.5 to 2. Such studies will not be confined to imaging and searching for the structural evidence of interactions, merging, strong star formation (for example, "starburst" galaxies), and active nuclear sources. They will also be used for spectroscopy of high-surface-brightness regions, again particularly in the near-infrared region where the resolution gains over the ground-based observations are complemented by a markedly lower background. For galaxies with a bright nucleus, a space-based facility is uniquely suited to determining red shifts, since the optimum aperture dictated by signal-to-noise ratio (S/N) considerations will be much smaller than can be used with typical ground-based seeing conditions. The optimum aperture can be smaller by up to an order of magnitude or more, depending on the light profile, i.e., on the type of galaxy and the degree of nuclear activity.

Star Formation. One of the major scientific puzzles of our time is that of star formation. It is a process involving remarkable amplification of density and violent interactions with a surrounding medium as the protostars enter their final phases of collapse.

Complex hydrodynamic and magnetohydrodynamic processes appear to play an important role. Difficult theoretical problems are complicated by lack of comprehensive data on the conditions in the protostellar objects and detailed knowledge of the mechanisms at work. Clearly it is a process of great complexity that will require a broadly based investigation in the radio (millimeter) and far-infrared wavelengths during the early phases, to detailed infrared and optical studies at the protostellar phase.

Adaptive-optic techniques will probably give ground-based telescopes diffraction-limited performance in the atmospheric windows in the 5- to 20-μm region, with speckle interferometry resulting in similar performance at shorter wavelengths. But the ability to image directly and to interpret unambiguously the data *at all wavelengths* from shortward of 0.4 μm to 10 μm at resolutions of 10 to 200 milliarsec would be of extraordinary scientific value. This is particularly true when the low background environment and the low emission from an ambiently cooled (100K or less) telescope are also considered.

The high-spatial-resolution infrared spectroscopic capability of such a telescope can be used during the intermediate phases of the collapse of a protostar, when its size is still near 1 arcsec (1000 AU in size at 1 kpc), to probe the circumstellar environment by means of spectral features. However, the most fundamental developments in this area will arise by investigating the properties of protostellar disks. Recently, using IRAS, we have discovered the existence of particles and dust orbiting nearby stars. We have also observed the first substellar companion of a nearby star. These disks have scales of 0.1 to 1 arcsec at 1 kiloparsec (kpc), ideally matched to resolutions of 40 to 100 milliarcsec over the 2- to 5-μm region. Detailed spectroscopic studies in the 2-μm region using the hydrogen molecule as a probe will reveal dynamical properties of the gaseous flows on scales as small as 50 AU at 1 kpc or 5 AU at 100 pc.

Stellar Systems. A large space telescope is uniquely suited to stellar astrophysical programs in the ultraviolet. In fact, the ultraviolet resolution would be better than 0.3 pc in M101 and less than 0.8 pc in Virgo. Thus, the whole area of spectroscopic studies of stellar populations in fields much too crowded for HST becomes possible. These include studies not only in the cores of globular clusters, in star-forming regions in the nearest galaxies, but also in the nuclei of galaxies—even in the Virgo cluster where young or ultraviolet bright stars could be identified and studied. Depending on the degree of confusion with background sources, it should be possible to image and construct color-magnitude diagrams for the old giant stars in the outer regions of Virgo cluster galaxies. Limiting magnitudes will exceed 31. Such a limit will allow the ages of globular clusters in M31 to be determined. It will also allow us to ascertain the age of M31 and that of the dwarf galaxies in the Local Group.

With resolutions of 10 milliarcsec or less, the capability exists for determining transverse motions within nearby galaxies and groups and even the nearest clusters. A critical factor here is the reference frame. While quasars may be too scarce in most fields, even with the several-arcminute field possible in such a telescope, the nuclei or any compact structure within background galaxies will form a satisfactory reference frame. Given such a reference frame, the limitation becomes the signal-to-noise ratio attainable and the precision to which the detector's spatial uniformity can be mapped. With adequate sampling and current ground-based CCDs, images can be routinely centroided to 0.01 of a full width at half maximum (FWHM), with measurement to 0.001 of a FWHM having been demonstrated. For unresolved sources, this latter value corresponds to being able to measure the positions of objects to a precision of 10 μarcsec. Distances to stars could be measured directly by parallax throughout our galaxy to distances of 10,000 pc. A yearly proper motion of this amount would result from a transverse velocity of only 30 km/s at the distance of M31 or M33, 300 km/s at M101, and 750 km/s at the Virgo cluster.

An exciting consequence of this capability is that the distance to a nearby galaxy could be measured independent of all other steps in the distance scale. For objects in circular orbits, the transverse motion can be compared with their rotational velocity to give a distance. The biggest uncertainty would be in the inclination of the galaxy.

Individual Stars and Binary Systems. At resolutions approaching 10 to 40 milliarcsec (wavelength 0.5 to 2 μm) some of the nearest giants and many more binaries can be resolved by a large space telescope, without the ambiguities associated with the interpretation of speckle interferometry. Such direct observations can be used to identify structures on stars and to investigate circumstellar shells of outflowing gas and dust in OB stars, Mira variables, and carbon stars. They will even allow the monitoring of atmospheric changes during pulsations of a few late-type stars, namely K and M giants and supergiants. In addition, many spectroscopic binaries can be resolved by such a telescope, adding dramatically to the number of stars for which stellar masses can be determined.

Planets. One of the most outstanding achievements of the space program has been our ability to explore the solar system, particularly for signs of life. The proposed telescope will allow such exploration to move out to the nearest stars, where Earthlike planets will be detectable if they exist. If such planets are found, the telescope will be powerful enough to detect life on them if it is like that on Earth. A large telescope as proposed is crucial for this study. The direct observation of the nearest star's planets, even if they are as large as Jupiter, will remain beyond or at the very limit of the HST. Visible light imaging requires surfaces of exquisite smoothness to avoid the scattering of the starlight that would overwhelm the planet's image, some 10^9 times weaker. In the thermal infrared, where the contrast is better, neither the HST or LDR will have the diffraction-limited resolution or be cold enough to be useful.

In this domain the large telescope could bring extraordinary new power. It could image and study by spectroscopy the atmospheric composition of not only Jupiterlike but Earthlike planets of nearby stars. It would be possible to search for oxygen, whose presence would be of extraordinary importance; the oxygen and ozone molecules in the terrestrial atmospheres originate in living organisms.

The nearest single stars similar to the Sun are at a distance of 4 pc. An Earthlike planet at 1 AU radius would lie at 1/4 arcsec from the star and could be resolved at 10 μm, where there is a strong ozone band, by a 16-m telescope. Cooling to 80K would bring the telescope emission below the zodiacal background, and a surface quality like that of HST would keep scattered light to the same low level. With these properties, Earthlike planets would be detectable, and the 10-μm ozone band could be measured in a few hours of integration.

The telescope will also be extremely valuable for studies of our own solar system. While sharing in the excitement of the planetary encounters such as that of Voyager with Uranus recently, the task group is concerned that the questions and theories resulting from the data from such missions must remain unanswered and untested for many years, if not decades, until the next mission. A telescope of the size and resolution suggested could do much to fill in these intervals and to allow planetary research to progress smoothly. For example, resolutions of 10 milliarcsec correspond to 35 km at the distance of Jupiter and 70 km at Saturn, allowing not only detailed imaging and spectroscopic studies of the planets themselves, but also spectroscopic analyses of volcanoes and other geological features on their moons. Such resolution is also invaluable for imaging and spectroscopic studies of Saturn's rings.

Technological Developments

It is clear that meeting the scientific goals outlined here for the use of an 8- to 16-m telescope will require imaging array detectors of the highest capability in the ultraviolet, optical, and infrared regions. They must have high quantum efficiency, very low noise, wide wavelength response, and be packageable into very large mosaic formats without compromising their performance. Technology development in this area is crucial if the value of the telescope is not to be diminished by less-than-optimum detectors. While challenging, development of such detectors is plausible and, in fact, possible even now through a phased research and development program.

Turning to the telescope itself, it is clear that the technology required to build an 8- to 16-m telescope is advancing rapidly. The 10-m Keck telescope will be made from 36 segments mounted together as one mirror. A number of other ground-based telescopes using 8-m monolithic mirrors are also planned. Methods for polishing diffraction-limited large monoliths or segments are now under development. Two possible avenues are available for orbiting the large telescope. It seems likely that in the time frame under consideration large vehicles could launch a prefabricated telescope up to 8 m in diameter. Alternatively, for a 16-m diameter telescope, construction in orbit is probably the best route. Mirror segments would be polished and tested on the ground and assembled onto a frame structure built in space.

Large telescopes designed to operate in a zero-g environment, but which do not have to withstand launch, are an exciting challenge to designers and engineers. Given a well-directed technology development program, the task group anticipates that an 8- to 16-m telescope will prove to be within closer reach than a simple extrapolation from HST would suggest.

VERY HIGH THROUGHPUT FACILITY (VHTF)

Although AXAF will be the first major observatory for x-ray astronomy with both sensitivity and resolution comparable with the most sensitive optical and radio facilities, it will be limited in its capabilities for spectroscopy of faint objects as well as for high-time-resolution studies owing to its relatively modest collecting area. The European x-ray project (XMM), will complement AXAF with a larger effective area but lower resolution so that fainter diffuse sources can be studied spectroscopically. Major extensions of this spectroscopic capability are crucial for addressing a broad range of fundamental problems in astrophysics. A Very High Throughput Facility (VHTF) would provide the required high-sensitivity spectroscopy as well as high-time-resolution studies of faint sources. The key to the VHTF is very large collecting area, possibly at the expense of angular resolution for spectroscopy and time variability studies. The VHTF, which would be assembled on a space platform with support and servicing from the Space Station, would consist of a grazing-incidence telescope system with total effective area of about 30 m^2. It would be constructed as either a single mirror of very large diameter and focal length, or more probably as an array of smaller telescopes of more compact design.

With this sensitivity increase, a number of qualitatively new investigations are possible, including the following:

- Dark matter in galaxies and clusters. VHTF would, with its enormous sensitivity for imaging and spectroscopy of diffuse objects, allow halos of galaxies to be measured for their total content of low-mass stars, diffuse hot gas, and total gravitational potential (by spatially resolved studies of its hot gas). Similar studies of galaxy clusters out to moderate red shifts ($Z \sim 0.5$) would allow temperature, density, composition, and mass profiles to be derived. This would constrain the still uncertain theories for the origin and evolution of hot gas in clusters.

- Star formation in molecular clouds. VHTF would image and locate pre-main-sequence stars, already known from Einstein observations to be relatively luminous x-ray sources in dark clouds. When coupled with deep infrared observations of star formation sites, physical conditions could be derived. X-ray heating of the cloud by its pre-main-sequence stellar population appears to play a fundamental role in the physics of cloud collapse and star formation.

- High-time-resolution studies of compact objects. VHTF would provide the ultimate capability to explore the physics of compact objects, accretion disks, and extreme field conditions in astronomical objects. With its imaging advantages, high spectral resolution (about 10^3 to 10^5), and large area it would study compact objects in our galaxy and nearby galaxies of the Local Group in great detail as well as QSOs at the largest red shifts. For example, through time-resolved spectra of x-ray bursts from a larger fraction of the burst sources in M31 and other nearby galaxies (observable within a single VHTF field), the mass and radii of neutron stars can be derived and compared with similar results for objects in our own galaxy. Detailed timing studies of galactic bulge x-ray sources in our galaxy could detect pulsations at a level of 10^{-4} of the persistent flux. This could detect stable pulsation periods and thus enable searches for the gravitational waves expected if the sources have very fast millisecond spin periods. High-resolution spectra of QSOs and distant galaxy clusters would measure red shifts directly from their iron-line features as well as probe the internal dynamics of accretion disks and jets where thermal (line) components are expected.

The category of large throughput x-ray instruments could also include a very large area array of proportional counters provided with mechanical collimators (about 1 degree revolution) solely to isolate relatively bright sources. A potential design goal would have an effective aperture of 100 m^2, sensitivity from about 0.2 to 40 keV, and timing resolution down to a few microseconds. Such an instrument would allow extreme phenomena in the vicinity of neutron stars and stellar-mass black holes to be probed in detail. The broad energy bandwidth would be vital in studying regions of high opacity. Such a large array could be built in space in modular form and assembled as a relatively low-cost experiment at the Space Station. This nonimaging detector would complement both the imaging soft/medium x-ray facility (VHIF) and a possible Hard X-ray Imaging Facility (HXIF).

HARD X-RAY IMAGING FACILITY (HXIF)

At energies above 20 keV, grazing-incidence x-ray telescopes are impractical, and this band will still remain relatively unexplored in 1995. The modest-sized (about 10^3 cm^2) hard x-ray detectors planned for the Franco-Soviet SIGMA Satellite (1988) and U.S. XTE mission (1992) should have made significant advances in detecting the brightest several hundred sources in the 20- to 200-keV band by that time, but detailed astrophysical measurements and exploration of the full hard x-ray/soft gamma-ray energy band of about 20 keV to 2 MeV will not yet have been possible. This energy range contains a rich assortment of information that can be used to address each of the field's three major scientific objectives: the early universe, compact objects and stellar collapse, and star formation. It is vital that we study this gap in the electromagnetic spectrum in detail.

A Hard X-ray Imaging Facility (HXIF) would provide a large increase in effective area (by a factor of about 300) and therefore an increase in sensitivity over any hard x-ray experiment flown previously. It would employ coded-aperture and Fourier-transform imaging with 10 arcsec to 1 arcmin resolution in a 5° field of view over the broad energy range up to a few million electron volts. Systems with very long effective focal lengths (10 to 100 km) between the coded mask and position-sensitive detector could achieve milliarcsecond angular resolution. Coded-aperture imaging techniques, using perforated occulting aperture plates (50 percent open area) to cast a shadow on a position-sensitive hard x-ray detector whose output is correlated with the mask, should have been fully developed and tested in flight (including the SIGMA mission) by 1995. With the large sensitivity increase possible with HXIF, imaging is essential in order to eliminate source confusion. It is also possible that direct (true) hard x-ray imaging over a more limited energy range can be achieved with, for example, Bragg concentrators. New approaches to hard x-ray imaging might be developed with a vigorous program of flight opportunities for low-cost experiments from the STS and Space Station.

HXIF could consist of an array of relatively simple (and self-contained) but large imaging telescopes, each with coded mask, shielded detector (probably scintillation crystals), and position-sensitive readout. The full array could consist of 64 modules, each with a 0.5 m × 0.5 m detector and mask at a focal length of 3 m.

(A separated detector/mask system with much higher resolution might eventually be operated at a lunar base.) This would yield a total effective detection area (through the mask) of 16 m^2 and would occupy a total volume of perhaps 5 m × 5 m × 4 m. The entire array would be co-aligned and fixed to about 1-arcmin accuracy and then pointed with about 10-arcsec stability.

The sensitivity of HXIF would be at least 200 times that of the experiments flown thus far and at least 10 to 30 times that of SIGMA or the X-Ray Timing Explorer Satellite (XTE). As such, it will be possible to attack a range of fundamental problems including the following:

- Central engines of quasars. QSOs and active galactic nuclei radiate most of their energy in the hard x-ray band. The 50 active galactic nuclei, for which spectra were measured out to about 20 to 50 keV with the HEAO missions, show relatively similar power law spectra with a spectral index of 0.7. If these spectra continue unbroken out to about 1 MeV, the total contribution of all active galactic nuclei would greatly exceed the hard x-ray/soft gamma-ray background. The total luminosity (determined by the break in the high-energy spectrum) of these sources is at present unknown. HXIF would be sensitive enough so that with a 10^4-s observation it should always detect and precisely locate one or two sources in its field of view and measure their spectra out to at least 300 keV. At 100 keV, the sensitivity would be sufficient to measure about 10 sources in each field.

Thus, it will be possible to measure the total energy output of QSOs for the first time. It would also be possible to measure changes in spectra and total luminosity as a function of cosmic epoch (red shift). With the sensitivity to observe about eighteenth-magnitude QSOs, HXIF could measure the brightest QSO at red shifts of $Z = 2.5$ and a broad range of luminosities at $Z = 0.5$. Very long exposures could reach correspondingly deeper (to $Z = 3.5$) so that the full spectra of QSOs at the earliest epochs could be probed.

- Physical properties of neutron stars and black holes. The sensitivity of HXIF would be sufficient to detect (at about 100 keV) gamma-ray burst sources in M31. Similar observations of the Magellanic clouds would yield accurate source locations (about 10 arcsec) and time-resolved spectra for each burst so that their hypothetical association with very low-accretion-powered neutron stars could be studied in detail. Cyclotron line features would be detectable and neutron star magnetic fields measured. Annihilation line features also could be studied in detail for burst sources in our galaxy so that gravitational red shifts for neutron stars could be measured.

High-time-resolution (less than 1 ms) studies at 300 keV of stellar-mass black hole candidates such as Cyg X-1 could be carried out for the first time, allowing the conditions nearest the hole to be more completely specified than with soft x-ray ($\sim$10-keV) studies alone. Comptonization studies versus time (in flares) in both galactic and extragalactic candidate black holes would allow the electron densities and temperature profiles to be derived and the physical size of the sources to be measured.

GAMMA-RAY ASTRONOMY

Following the Gamma Ray Observatory (GRO), several telescopes will again be required to meet the objectives of gamma-ray astrophysics because of the different interaction processes involved in gamma-ray detection over the large gamma-ray energy range, 10^5 to 10^{11} eV. Even with the relatively large instruments on the GRO, gamma-ray astronomy is constrained by the number of detected photons. Larger-area telescopes with longer exposures and markedly improved angular resolution are necessary to meet the objectives beyond GRO. Energy measurements are important over the entire spectrum, with the required resolution depending on the energy interval.

Gamma-ray observations are particularly relevant for those phenomena in which high-energy processes reflect the underlying energetics of the system. Active galactic nuclei, compact objects, explosive phenomena, and the acceleration and interactions of cosmic rays are examples. We must obtain accurate measurements of the gamma-ray luminosity of large numbers of active galactic nuclei so that different production mechanisms can be identified by class and their contribution to the diffuse radiation can be determined accurately. Temporal variability of these sources on time scales from minutes to years will be required to distinguish whether the high-energy emission arises from a central engine (e.g., a massive black hole), from interactions of energetic jets with the ambient material in the sources, or as the result of individual explosive events (e.g., supernovae). Observations extending over many years will be required to interpret accurately the relationship of the central energy source to phenomena observed at other wavelengths, such as the relativistic jets and superluminal expansion of knots. For some sources, we can utilize the temporal variability to associate conclusively the gamma-ray object with observations at other wavelengths. In general, however, gamma-ray detectors with source location capabilities approaching 1 arcmin will be necessary.

The detailed study of compact objects—black holes, neutron stars, dwarf stars—in our galaxy and nearby galaxies will demand high-quality measurements. Nuclear line emission resulting from reactions of energetic particles with the surface material of neutron stars should provide direct information about composition of this material and of surface red shifts from which the mass-to-radius ratios can be derived. With estimated line fluxes from 10^{-5} to 10^{-7} γ /cm^2/s even for relatively nearby objects, substantial sensitivity improvements beyond GRO are required to address these studies satisfactorily. Spectral and temporal characteristics of compact sources will also allow model-dependent probes of the inner portions of the accretion disks. The dynamical and spectral characteristics will provide additional information on those sources that contain black holes, and thereby provide tests for physical processes occurring in the vicinity of black holes. The study of explosive events, supernovae and novae, will benefit from the substantial improvements in sensitivity and spectral resolution anticipated with the instruments in the 1995 to 2015 era. Significant data on extragalactic supernovae will provide direct, quantitative information on the acceleration of cosmic rays and nucleosynthesis of heavy elements.

Currently, the following instruments appear necessary for progress in gamma-ray astronomy:

1. State-of-the-art spectroscopy in the 0.1- to 10-MeV spectral region for nuclear gamma-ray line observation can currently be accomplished with solid-state detectors, e.g., germanium detectors. At present, efforts are under way to develop large germanium arrays that could provide the desired sensitivity—significantly below 10^{-5} γ /cm^2/s—if background problems can be overcome. Position sensitivity within germanium detectors is being pursued, and a large array combined with a coded mask could provide a system that combines high spectral and angular resolution. The possible

use of massively shielded detectors for use in this low-energy region will also be considered. Such an instrument should be able to observe radioactivity in supernova remnants, electron-position annihilation, and nuclear excitations caused by cosmic rays.

2. In the medium-energy gamma-ray range (1 to 60 MeV), an advanced Compton telescope where both upper and lower detector elements have high energy and spatial resolution and low background would result in considerable improvements in sensitivity and energy resolution compared with previous instruments. Good energy resolution in the million electron volt range will permit an accurate study of the spectra of individual sources, thereby permitting a better understanding of their origin.

3. In the gamma-ray region above 50 MeV, where the basic pair-production processes provide an inherently low background, much greater sensitivity (about an order of magnitude) will come from increased area and improved efficiency and angular resolution. Angular resolution approaching 1 arcmin is also required to achieve the desired point-source location. Detector systems under development, including large, high-position-accuracy particle-location chambers, combined with new large telescope designs, should provide the desired improvements. Such an instrument would search for faint objects and make detailed studies (including time resolution) of galactic sources. It would also dramatically extend the knowledge of high-energy phenomena of active galactic nuclei. Extending the energy range upward ($E > 10^{11}$ eV) will be important in understanding the origin and acceleration of the high-energy cosmic rays.

The approaches to these instruments seem feasible, but development of the new generation of detectors must be funded at an adequate level now to ensure that advanced instruments will be available in the 1995 to 2015 era.

COSMIC-RAY RESEARCH

New programs in particle astrophysics will explore energy regions far beyond those currently accessible and will probe the particle population in our galaxy at greatly improved levels of sensitivity and resolution. Specific goals for cosmic-ray research are precise measurements of the isotopic abundances at energies (~10 to 100 GeV/nucleon) that are well above the region of solar modulation; measurements of the energy spectra and isotopic abundances of ultraheavy particles; determination of the spectra of electrons, positrons, and antiprotons over a wide energy range; sensitive searches for heavy antiparticles; and measurements of the composition of extremely energetic particles (well beyond the "bend" at 10^{15} eV). Realizing these goals will enable us to test and specify theoretical models on the acceleration of particles in our galaxy and beyond, to analyze key evidence for the nucleosynthesis of the elements, to study structure and composition of the interstellar medium, and to provide observational tests for cosmological models. A large portion of the observational program is expected to be centered around a magnet Spectrometer Facility (ASTROMAG) that should be in orbit in the early 1990s. In addition, very large detector arrays should be assembled in near-Earth orbit to detect expectionally rare but important particle species. Polar-orbiting platforms should serve for investigations at low energies. A particular challenge are measurements on a deep-space probe reaching interstellar space outside the heliosphere for detailed in situ investigation of the interstellar medium.

Magnet Spectrometer for Particle Astrophysics (ASTROMAG)

ASTROMAG will be a superconducting magnet spectrometer exhibiting field integrals of several teslameters over an area of at least 1 m^2, combined with large-area, trajectory-determining devices with better than 100-μm resolution. Such an instrument permits precise measurements of the rigidities of high-energy particles, far beyond the capabilities of conventional instrumentation. This spectrometer is expected to be in earth orbit for a duration of 10 to 20 years and many require occasional servicing. To perform specific astrophysical observations, the spectrometer must be combined with dedicated particle-detector systems. For instance, isotopic abundance measurements require an accurate velocity measurement with Cerenkov counters, in addition to the rigidity measurement by the magnet spectrometer. On the other hand, the identification of singly charged particles (protons, antiprotons, electrons, and positrons) must be accomplished by complementing the spectrometer with transition radiation and shower detectors. Not all observations can be performed simultaneously, but the dedicated detector systems should be successively accommodated and interchanged, like focal plane instruments on a telescope. The following briefly describes some of the observational objectives:

- *Antiprotons and Antimatter Search*: to search for antiparticles in the cosmic rays, e.g., antiprotons, heavy antinuclei, and positrons. Such observations relate to the fundamental question of matter-antimatter symmetry of the universe, the missing-mass problem in cosmology, and the production of antiprotons and positrons by interactions in interstellar space.
- *Isotopic Composition of High-Energy Nuclei*: to measure the isotopic composition. This will address questions of nucleosynthesis, origin of the elements, and evolution of the galaxy. It will also address the problem of dating with radioactive isotopes, and the study of interactions of cosmic rays with the interstellar gas.
- *Energy Spectra*: to determine precise energy spectra of cosmic rays over a large energy range in order to understand the processes of particle acceleration on astrophysical scales and of the confinement of cosmic rays to our galaxy.
- *Electrons and Positrons*: to measure negative and positive electrons up to energies of a few tesla electron volts. High-energy electrons reaching the Earth cannot have traveled large galactic distances; thus the details of their energy spectra will reveal information on the spatial distribution of acceleration sites in our galaxy.

Interplanetary and Interstellar Measurements

In situ measurements of low-energy cosmic rays will be performed with several detectors on spacecraft or space probes at different locations throughout the heliosphere. These detectors will use proven solid-state detector technology, or they may employ new devices now in development that permit much larger detector areas, with a corresponding increase in sensitivity. These instruments will study the three particle populations in the heliosphere: galactic cosmic rays, energetic particles of solar or planetary origin, and the anomalous cosmic-ray component. These phenomena will be studied at detection levels that permit investigations of very rare species, including ultraheavy particles, and with high-mass resolution sufficient to identify rare isotopes (a related role will be played by detectors on a polar platform; see the section below on Experiments on Polar Platforms).

One or several space probes will reach nearby interstellar space, outside the region of solar modulation. A dedicated interstellar probe will make it possible to pursue one of the most important and most challenging goals in the coming decades: in situ measurements in interstellar space. Recent measurements on the Pioneer and Voyager spacecraft have dramatically expanded our understanding of the solar environment. The size of the heliosphere, estimated to be 5 to 10 AU before these missions, is now estimated to be 50 to 100 AU in the ecliptic plane.

Large Detector Arrays in Space

Some of the most critical questions in particle astrophysics will only be answered by the exposure of arrays of detectors larger or more massive than those that can be carried on a single Shuttle mission. These arrays will have to be assembled in space from separate modules carried up individually. The following are examples:

- *High-Energy Array*: 10^{15} to 10^{16} eV. The deployment of an exceptionally massive array is required in order to study cosmic-ray particles in the energy range beyond 10^{15} eV, where air-shower data suggest a break in the energy spectrum. This break might reflect the large-scale structure of the galaxy and the escape of high-energy cosmic rays from the galactic magnetic fields, or it may signify limitations in the galactic acceleration mechanism. To answer these questions, precise measurements of the elemental abundances are necessary. This will require an array that is large enough to detect a significant number of particles at these high energies and is able to measure their energy. Only a large calorimeter appears to meet these requirements. Such a calorimeter would have a mass of 60 tons and a diameter of 5 m, or even more.

- *Ultraheavy Nuclei*: $Z > 30$. For the very rare ultraheavy (UH) nuclei, individual elemental abundances will be determined with the Heavy Nucleus Collector (HNC) (see Chapter 3), but practically nothing will be known about their energy spectra. Only by achieving major increases in the collecting power of the detectors will it be possible to acquire the data needed. Such data will not merely supplement what we already know but will add a new dimension because of the different histories of nucleosynthesis and propagation of these heavier elements. An array has been proposed that is composed of a large number of relatively simple detectors, arranged in a sphere with a diameter of 30 m and a mass of 30 tons.

Experiments on Polar Platforms

Exposure of large instruments in high-inclination orbit for extended periods of time is required to address several remaining frontiers of cosmic-ray research. Examples include measurements of cosmic-ray positrons and antiprotons from 0.1 to approximately 4 GeV and measurements of the isotopic composition of ultraheavy nuclei ($Z > 30$). The ultraheavy isotope measurements are considered in more detail below.

For those elements occurring beyond iron and nickel in the periodic table, elemental composition measurements will provide partial information on the nucleosynthesis, acceleration, and transport of ultraheavy cosmic rays in the galaxy. Further details can be deduced from the abundances of individual isotopes. Similarly, studies of the time history of ultraheavy cosmic rays, from their nucleosynthesis and acceleration through propagation throughout the galaxy, will require the determination of the abundances of individual radioactive isotopes.

The investigation of the isotopic composition of the ultraheavy elements requires the achievement of good mass resolution in large-area detectors. It also requires the availability of long-duration (several years) exposures of such instruments to the galactic cosmic-ray flux. Typical detectors will require a very large area (several square meters) and will most likely be restricted to low energies (< 1 GeV/n). Because of the geomagnetic cutoff, these measurements cannot be performed aboard a Space Station in low-altitude/low-inclination orbit. The ideal vehicle would be capable of getting out of the magnetosphere. However, since the instruments to be flown will necessarily be large and heavy, access to such an orbit may not easily be available. A satisfactory alternative would be a near-Earth platform in a near-polar orbit. Even at the lowest energies of interest for galactic cosmic-ray studies (around 50 MeV/AMU), data will be collected over approximately one third of each orbit.

5
Practical Considerations

The task group believes that the program outlined above is visionary in its scientific implications, and yet remains feasible. There are, of course, a multitude of budgetary, management, operational, and technological issues that will have to be resolved in order to execute this program. Below, the task group has attempted to assess the most important of these issues using the projected program of Chapter 4 as a base.

BUDGETARY REQUIREMENTS

The task group has attempted to estimate the financial resources necessary to accomplish this program. The basis for its estimate is straightforward:

- The task group assumes that the Support Research and Technology (SR&T) and Explorer-type programs will be maintained at about the current level (about $100 million per year). Every effort should be made to increase these budgets because the future initiatives proposed in this report may depend critically on developing new technologies and approaches to conducting experiments in space.
- In addition, the task group takes the cost of developing, maintaining, servicing, replacing, and upgrading the current or

planned observatory-level facilities to correspond to the recommendations made by the NRC's Astronomy Survey Committee in 1982 at current budgetary estimates.

• For each major wavelength regime, the task group assumes that one and only one observatory-class facility will operate at a time.

• The task group has assumed that in each wavelength band a new facility will be developed soon enough to prevent the occurrence of substantial gaps in observational capabilities due to facility deterioration or obsolescence.

• In all cases the task group assumed that the old facility is replaced by a new one at about twice the cost of the facility that is replaced.

• The cost of new initiatives such as optical interferometry and radio interferometry has been estimated to be comparable with that of current observatory-class missions.

The task group considers its estimates to be on the high rather than low side if more of the expected efficiencies in development and operation should materialize.

The task group concludes that the base program (Astronomy Survey Committee recommendations and replacements plus SR&T and Explorers) could be carried out with a modest annual real-dollar increase of the NASA fiscal year 1985 astrophysics budget at 2.3 percent per year. By 2015 this budget would be $1 billion per year. Such a program would mainly utilize the Space Station capabilities for service and maintenance of co-orbiting platforms and free flyers.

A further increase in the rate of growth of the current fiscal year 1985 NASA astrophysics budget at 3.7 percent per year would be required to pursue vigorously the new opportunities afforded by the Space Station for in-orbit assembly and deployment of large structures. The budget for astrophysics would then be of the order of $1.5 billion per year by 2015.

INTERNATIONAL COLLABORATION

International collaboration can occur on at least two levels. On one level scientists from other nations may collaborate as part of a mission team. Alternatively, they might compete by providing their own proposals for space on a U.S. satellite. This level of collaboration has provided the United States with an important source of new ideas and challenges. It has also provided the United States with some leverage to influence the development of international space science.

At a different level, a joint effort can be envisaged—as has occurred in the past—where the flight hardware of a whole program or mission is produced as the result of an international effort. Past experience shows that such undertakings are successful only if a number of prerequisites are met:

- The parties must be competitive in order to be of interest to each other.
- Joint planning must occur at an early stage.
- Clear goals and managerial procedures must be defined.
- Continuing support of the project by the parties must be assured.

The United States now has competitive partners, mainly in Japan and Europe. Further collaboration on major projects will require joint planning. Collaborating nations will have to become accustomed to the idea that leadership on any one project can rest with any one of the parties. The task group envisages a system where missions are managed by one participant, and yet scientific instrumentation and exploitation are open to all.

COST-TO-WEIGHT RATIO

The cost of a space mission seems to be directly proportional to its weight. Since the facilities being considered for future research are extremely large and heavy, their costs will be high unless we can develop ways to reduce the cost-to-weight ratio across the board in space research. Further, the mass of future instruments could be reduced through the use of sensors and active control systems for aligning large structures to optical tolerances.

MANAGEMENT AND OPERATIONS

The future research facilities that we are planning are expected to be long-lived, refurbishable while in orbit, and available to the whole astronomical community. The way that these facilities are operated and maintained is of crucial importance. We must study and gain improved understanding of different management approaches (e.g., principal investigators, guest observers, institutes), the costs of operating for long durations, scientific strategies for instrument replacement, and usable lifetimes for observatories within some budget constraints.

COORDINATED FACILITIES

Real advantages could accrue from the ability of several observatory-class instruments to share support and refurbishment capability. The advantages and disadvantages of a variety of such arrangements should be studied.

SCIENTIFIC INSTRUMENTS TECHNOLOGY

Interferometry

The need for higher angular resolution, with a microarcsecond as a target, will require examination of the present and projected state of interferometry techniques from radio wavelengths to the ultraviolet. Technical needs of interferometry can be identified in three areas: structural technology, optical technology, and station keeping. Among the subjects needing study in the area of structural technology are the construction, measurement, and control of large precision structures; the precision of control of pointing and momentum exchange; vibration minimization and decoupling; and the general area of metrology for high-precision monitoring of the structures. Optical technology is a central concern, including the development of active systems, sensors, fiber optics (especially single mode), and the study of image reconstruction. Finally, station-keeping technology needs further study, including precision position and altitude control, quiet thrusters, orbital analysis, and studies of improved contamination control. Finally, a study is needed of methods of reconstructing images and the closely related question of fringe detection, both as they relate to detectors per se and as they relate to algorithms for tracking fringes.

A sequence of missions needs to be defined to lead to the achievement of the above goals for resolution. Most likely the early progression will emphasize small instruments. These could include a quasi-imaging device with 1- to 10-milliarcsecond resolution and incomplete ultraviolet plane coverage. Eventually, larger instruments with higher collecting areas, greater resolution, and complete ultraviolet plane coverage will be possible, based on the experience with smaller instruments.

Detectors and Techniques

A variety of new detectors and techniques for possible use in cosmic x-ray astronomy and astrophysics are likely to be developed over the next 10 to 15 years and would thus be of interest for the new initiatives described in Chapter 4 in the sections on the Very High Throughput Facility and Hard X-ray Imaging Facility. A particularly exciting development is the x-ray bolometer detector/spectrometer, now being developed as a possible focal-plane instrument for AXAF. This should allow energy resolution of a few electron volts over a broad bandwidth (0.1 to 10 keV) with nearly unit quantum efficiency by making use of cooled (to helium-3 temperatures) silicon or germanium bolometers in which the thermal energy associated with absorption of individual x-ray photons is detected. Small detectors, with limited imaging arrays, are a likely next step for follow-on facilities such as the VHTF. New detector systems with modest energy resolution but very high spatial resolution will also likely become available for both the soft and hard x-ray bands.

New techniques for x-ray imaging are being developed for both soft and hard x rays. Two approaches for achieving the very large collection areas required for a VHTF system are being developed for evaluation on Shuttle flights of short duration. One employs thin foils to make highly nested arrays of telescope mirrors. The other consists of relatively compact arrays of curved glass plates to be assembled in modular arrays of telescopes. At energies below about 1 keV, narrow-bandwidth imaging telescopes operating at nearly normal incidence are being developed for solar applications; scaled-up versions of these could be used for high-sensitivity spectral line imaging of cosmic x-ray sources. At hard x-ray energies (above 10 to 30 keV, where grazing-incidence optics become impractical), coded-aperture and Fourier transform imaging techniques are being developed on balloon and Shuttle experiments, and could be applied to the HXIF. A longer-range program to develop direct imaging using Bragg concentrators at energies up to about 100 keV should also be encouraged. Finally, the development of x-ray lasers is now being pursued for nonastronomical purposes. If successful, the applications to future x-ray missions are very promising, from heterodyne interferometry to high-resolution spectroscopy.